CW01655170

£4

Find, Fix and Strike!

The Fleet Air Arm at War 1939–45

John Winton

WITHDRAWN
SOLD

B. T. BATSFORD LTD, LONDON

A PROPHECY

The time will come, when thou shalt lift thine eyes
To watch a long-drawn battle in the skies,
While aged peasants, too amazed for words,
Stare at the flying fleets of wondrous birds.
England, so long the mistress of the sea,
Where winds and waves confess her sovereignty,
Her ancient triumphs yet on high shall bear,
And reign, the sovereign of the conquered air.

From *Luna Habitabilis* (That the Moon is Inhabitable)
by Thomas Gray, 1738

First published 1980
© John Winton 1980

All rights reserved. No part of this publication
may be reproduced in any form or by any means
without permission from the publishers.

ISBN 0 7134 3488 0

Photoset, printed and bound
in Great Britain by
REDWOOD BURN LIMITED
Trowbridge & Esher
for the publishers B. T. Batsford Ltd,
4 Fitzhardinge Street, London W1H 0AH

Contents

List of Illustrations

BLACK AND WHITE PHOTOGRAPHS

MAPS

Sources and Acknowledgments

Grateful thanks and acknowledgments for permission to publish are due to Cdr Charles Lamb, DSO DSC RN, for extracts from *War in a Stringbag* (Cassell, 1977); to Captain Godfrey French, CBE RN, for an extract from *Some Operations of HMS Furious and Her Aircraft 1939–40* (*The Naval Review*, Vol. L, No. 1, January 1962); to Major Richard Partridge, DSO, Royal Marines, for an extract from his forthcoming book *Operation Skua*; to Cdr C. A. Jenkins, RN, for an extract from *Days of a Dogsbody* (Harrap, 1946); to Cdr Ronnie Hay, DSO DSC RN, for an extract from a letter to the author of 18 May 1978; to Joseph P. Lawrenson, for an extract from 'The Cruiser *Curlew*', from *70 True Stories of the Second World War* (Odhams Press), first published in *The People*; to Frederick Muller Ltd, for an extract from *Adventure Glorious*, by Ronald Healiss, 1955; to the Controller of Her Majesty's Stationery Office, for extracts from *Fleet Air Arm*, 1943, from Vol. II of *The War at Sea 1939–1945*, by Captain Stephen W. Roskill, 1957, from 'The Carrier-borne Aircraft Attack on Oil Refineries in the Palembang (Sumatra) Area in January 1945', by Admiral Sir Arthur John Power, in Supplement to the *London Gazette* of 5 April 1950, and from 'The Contribution of the British Pacific Fleet to the Assault on Okinawa, 1945' by Admiral Sir Bruce Fraser, in Supplement to the *London Gazette* of 1 June 1948; to Blackwood's Magazine for extracts from 'A Taranto Diary' by Lt M.R. Maund, DSC RN, of No. 1553, March 1945; to William Kimber Ltd, for extracts from *Taranto*, by Don Newton and A. Cecil Hampshire (1959) and *Task Force 57*, by Peter C. Smith (1969); to the Fleet Air Arm TAG Association for extracts from 'The Knights of Malta' by Nat Gold, and 'Albacores in Action' (Anon), both from the Association's 25th Anniversary Publications of 1972; to Mrs Dorothea Pack, for extracts from the late Captain S. W. C. Pack's *The Battle of Matapan* (Batsford, 1961); to the Editor of *The Naval Review* for an extract from 'HMS *Argus* 1914–1941', by R. L. S. (Vol. XXXIII, No. 1, February 1945); to Lt-Cdr Michael Apps, RN, for extracts from *Send Her Victorious* (William Kimber, 1971); to Rear Admiral Percy Gick, CB DSC, for extracts from a recorded tape of 17 April 1978 on 'The Attack on *Bismarck* by 825 Squadron from *Victorious*'; to John Hoare, for extracts from *Tumult in the Clouds* (Michael Joseph, 1976); to C. E. Friend, for an extract from 'The Attack on *Bismarck*', published in *Hamptonians at War*, ed. by W. D. James, privately printed for Hampton Grammer School, 1947; to John Deane Potter, for extracts from *Fiasco: The Breakout of the German Battleships* (Heinemann, 1970); to John Farquharson Ltd, for an extract from *Ark Royal*, by Rear Admiral William Jameson (Hart Davis, 1957); to Hugh Popham, for extracts from *Sea Flight* (William Kimber, 1954); to Terence Horsley, for an extract from *Find, Fix and Strike* (Eyre & Spottiswoode, 1943); to Kenneth Poolman, for extracts from *Escort Carrier 1941–1945* (Ian Allan, 1972); to A. D. Peters, for extracts from *Escort Carrier*, by John Moore (Hutchinson, 1944); to John Eames, for an extract from a letter to the author of 12 February 1978; to F. D. Ommanney, for an extract from *Flat Top* (Longmans Green, 1945); to Sir Geoffrey Bates, for 'Fleet Fighter' by Olivia Bates (née Fitzroy), from Fleet Poetry Broadsheet No. 3, June 1945, published at Naval Headquarters, Colombo; to Norman Hanson, for an extract from Richard Abrams' *F4U Corsair at War* (Ian Allan, 1977); to the Editor of the *Royal Air Force Flying Review*, for an extract from 'Palembang', by Major Hideaki Inayama (Volume XV, No. 8); to the wardroom officers of HMS *Formidable*, for extracts from *A Formidable Commission* (Seeley, Service & Co., 1946); *The Forgotten Fleet*, by John Winton, was published by Michael Joseph, 1969.

1

1939 Early Days and Early Losses

Early Days

On 3 September 1939 the Royal Navy had a rather better Fleet Air Arm than the country or the Navy really deserved. Between the wars, the RAF had had control over maritime air power. Naval air had been, literally, the fleet air arm of the Royal Air Force. Full control over the Fleet Air Arm was not finally wrenched from the RAF by the Admiralty until July 1937. The long, drawn-out controversy over who should direct it had inhibited the expansion, design and training of the Fleet Air Arm. The Admiralty itself had been uncertain as to what kind of aircraft it needed. There was nobody to speak for the Fleet Air Arm in high places. There were no naval pilots in senior positions ashore or afloat, and no admiral with a flying background.

Although some officers, such as Admiral Sir Ernle Chatfield, First Sea Lord from 1933–8, had always been determined that one day the Navy would get back its air arm, by no means everybody in the Service agreed that air power was either necessary or desirable. In 1908 the Admiralty, in the curt way in which, eighty years earlier, it had dismissed the possibilities of steam propulsion, wrote to the Wright Brothers that 'Their Lordships are of the opinion that they [aeroplanes] would not be of any practical use to the Naval Service' and there were still not a few naval officers who clung to that opinion. Captain Bernard Acworth, a not uninfluential writer on naval affairs in the 1930s, weighed up the possibilities of air power at sea and concluded that 'in peacetime, it may be frankly admitted, seaplanes available for picnics, shooting parties, or as substitutes for captains' galleys when lying at anchor far from shore, would be fun.' In 1936, when the war clouds were clearly thickening, one Commander-in-Chief wrote '. . . my general impression is that we are heading for a very ambitious carrier programme and we may be rating the carrier's value in War too high, in view of the fact that its power of hitting and destroying the enemy is not in measuring distance of the power possessed by other vessels.'

Where the need for aircraft was, reluctantly, accepted, it was thought that their chief tactical use would be to slow down enemy ships so that the main battle fleet could then overhaul and dispatch the crippled survivors. The concept of the aircraft as a pure and simple striking weapon on its own, with no bounds or qualifications, had yet to be widely accepted. Similarly, there were doubts that high-speed, high-performance aircraft could land safely on carriers' flight decks. There was a belief that naval aircraft, of necessity, had to be of poorer performance than land-based aircraft. There was no properly designed, accurate dive-bombing sight (and, by corollary, no efficient anti-submarine bomb or depth-charge for Coastal Command).

As a result of such intense inter-Service squabbling, controversy, prejudice, hoary myth, ignorance, and misplaced confidence, the Navy entered the war with no high-performance, single-seater, monoplane fleet fighter. The standard fleet fighter was the Sea Gladiator, a biplane converted from RAF use, with a top speed of about 220 knots armed with four .303 machine guns. The fleet torpedo-spotter-reconnaissance aircraft was the three-seater Swordfish, first built as a private venture by the Fairey Aviation Company, with a top speed of about 120 knots (and that, as the crews used to say, was 'downhill'). Just coming into general fleet service was the two-seater fighter-dive-bomber, the Blackburn Skua, maximum speed of 220 knots, armed with four .303 machine guns. The first Skua squadron was No. 800, embarked in *Ark Royal* in November 1938. There were three Skua squadrons by the outbreak of war, when the total number of front-line naval aircraft in service was 232.

The aircraft carriers themselves were a mixed collection. *Furious*, at 22,450 tons, was once a large cruiser, armed with two 18-inch guns, designed for service in the Baltic. She was converted as an aircraft carrier during the First World War, when she established two of the earliest records in naval aviation. Squadron Commander E.H. Dunning, RNAS, made the first ever deck landing on board *Furious*, in a Sopwith Pup on 3 August 1917. *Furious* was also the first aircraft carrier to launch a major air attack. From her deck on 19 July 1918 seven Sopwith Camels flew off for the successful bombing raid on the Zeppelin base at Tondern, where L-54 and L-60 were destroyed. *Furious* was always a lucky ship and a most happy one. All who served in her remember her with the greatest affection. She had a complement of some thirty aircraft but in September 1939 she was carrying out deck-landing training in the Firth of Forth, with no particular squadrons allocated to her.

Glorious and *Courageous*, sister ships of *Furious*, but designed to have four 15-inch guns each, had also been converted in the 1920s,

Courageous being completed in 1928 and *Glorious* two years later. Both had official aircraft complements of forty-eight but seldom carried as many. When the war began, *Courageous* was based at Portland with the Home Fleet, with 811 and 822 Swordfish squadrons, of twelve aircraft each, embarked. *Glorious* was in the Mediterranean Fleet, with 802 Squadron of six Sea Gladiators, and three Swordfish squadrons, 812, 823 and 825, each with twelve aircraft.

The small carrier *Hermes*, at 10,850 tons, was the first ship to be built as a carrier from the outset. She could carry some twenty aircraft but began the war with 814 Squadron, of twelve Swordfish, transferred from *Ark Royal*. *Eagle*, of 22,000 tons, was yet another conversion, laid down as the Chilean dreadnought *Almirante Cochrane* in 1913, bought by the Admiralty in 1917 and commissioned as a carrier in 1920. She was on the China Station when war broke out, with eighteen Swordfish of two squadrons, 813 and 824, embarked.

The newest carrier in the Navy was *Ark Royal*, at 27,000 tons full load, first commissioned in November 1938. She was based at Scapa Flow with the Home Fleet, with eighteen Skuas, of 800 and 803 Squadrons, and thirty-six Swordfish of 810, 820 and 821, and one Walrus amphibian, all embarked.

There was also *Argus*, of 14,000 tons, another conversion, who was used as a training carrier, and a tender for the 'Queen Bee' radio-controlled target aircraft. She, too, was kindly remembered by those who served in her, not least for her superior accommodation, a relic of her original designed purpose as the Italian Lloyd Sabaudo Line ss *Conte Rosso*. Work stopped on her in 1914, she was bought in 1916 and converted in 1918. Finally, the Australian-built seaplane carrier *Albatross*, transferred from the Royal Australian Navy in 1938, was sent to Hastings, Freetown, in West Africa, with six Walrus embarked in October 1939.

The men, too, were a mixture. There were about 500 operational aircrew in the Fleet Air Arm in 1939. Of the pilots, some were career naval officers who had chosen to fly, just as their contemporaries had become gunnery or signals or navigating officers. They were largely the 'loners', and non-conformers, the 'salt-horse' non-specialists, who abhorred gunnery and gunnery officers, and who had decided they were unlikely to be promoted from any of the normal specialisations. As a body of men they had an attractive and necessary mixture of professional self-reliance and eccentric idiosyncrasy. Some had transferred from the RNR, some had entered after previous experience with civil airlines. A good many had served in the RAF. As late as 1940, when the Fleet Air Arm's own training programmes for aircrew and ground staff were in full swing, there were still some 2000 officers and men of the RAF serving with the Navy. In 1939, the Fleet Air Arm had naval air stations at Donibristle, Gosport, Lee-on-Solent and Worthy Down, Ford, near Littlehampton in Sussex, and Eastleigh, and building had started at Yeovilton and Arbroath. The Navy had a second-line strength of 190 aircraft. (By comparison, in 1945 the Navy had forty-five naval air stations and 2790 second line aircraft.)

In spite of all the uncertainties and difficulties, the carriers had been in commission for several years. The Navy had experience of operating aircraft in various climates and weather conditions. There was at least a core of fully trained and experienced aircrew and ground staff. Tactics, offensive and defensive, had been practised. Officers and men shared an accumulated stock of operational knowledge, useful wrinkles, and flying lore and mythology. Personalities had emerged, aircraft limitations had been explored. Some far-sighted officers had even begun to exchange opinions and imagine air power used at sea on a scale and with a devastating effectiveness never dreamed of before. In short, the Navy had, most fortunately, the nucleus of a professional, effective, world-wide, naval air arm, skilled enough to strike at the enemy, and numerous enough to train their successors of the future, if only its ships and men could somehow survive the shocks and disasters which always seemed inevitable in the early days of any war at sea.

Loss of *Courageous*

At first, survival did not seem at all likely. Amongst some vestigial remains of First World War thinking still lurking in some corridors of the Admiralty was the notion that convoys were 'defensive' and therefore by definition, undesirable. By extension of the same argument, aircraft carriers should be directed 'offensively'. The tactics took the form of a series of punitive lunges, like those of a kind of bold anti-submarine cavalry, by the two Home Fleet carriers *Ark Royal* and *Courageous*, to wherever U-boats had been reported. The presence of the carriers would, it was believed, distract the attention of the U-boats from the merchantmen and their precious cargoes, luring them into positions where the Swordfish could hunt and harry them and keep them down, so that they might be located by the destroyers' asdics (at that time a much overrated device) and then sunk.

Certainly, the U-boats' attention was distracted—towards the carriers themselves. *Ark Royal* narrowly survived a torpedo salvo from U-39 on 14 September. The U-boat was sunk by *Ark*'s escorting destroyers, which seemed to justify the theory, but it had been a close shave and there was worse to come. Late on 17 September, *Courageous* was off the south-west coast of Ireland and was at cruising stations. It was a warm autumn evening and there was still an atmosphere of peace-time on board. Officers were still changing for dinner. *Courageous* turned into wind to land on her Swordfish, which had been out looking for a merchantman which had reported being attacked by a U-boat. Two of her four escorting destroyers were away from her, for the same reason.

Courageous's turn unluckily brought her directly in the periscope sights of Lt-Cdr Schuhardt of U-29 who was at the end of his patrol and was at that moment on his way as a final operation to intercept a convoy reported by another U-boat. At 7.50 p.m. U-29 fired a salvo of three torpedoes at 3000 yards' range. Two hit *Courageous* on her port side. Sub-Lt (A) Charles Lamb, who had just flown on his Swordfish, the last aircraft ever to land on *Courageous*'s deck, years later described in *War in a Stringbag* that the impact was 'the like of which I had never imagined possible. If the core of the earth exploded, and the universe split from pole to pole, it could sound no worse. Every light went out immediately and the deck reared upwards, throwing me backwards, and the hot blast which followed tore at the skin on my face and plucked at my clothes. There was something Satanic about it, and unreal. In the sudden deathly silence which followed I knew that the ship had died.'

Courageous heeled further and further over to port and sank in less than twenty minutes. She was the first British warship to be sunk by enemy action in the Second World War and she took 518 of her people with her, including her captain, Captain W.T. Makeig-Jones, an officer of the old school, who believed that a captain was always responsible at all times for his ship and if he hazarded her, unintentionally or not, it was his duty to go with her. He stayed on the bridge, saluting the flag, as the ship went down.

After *Courageous*'s loss no more of these anti-submarine sorties by carriers were made—this lesson at least had been learned. But *Ark Royal* had already entered upon her two years of charmed life, for on 26 September she had another near escape. Early that day she was heading eastwards towards the Norwegian coast with a number of other ships, to escort home the damaged submarine *Spearfish*, when shadowing Dornier Do18 flying boats were sighted. Skuas were launched and one, piloted by Lt B.S. McEwen, with Acting PO B.M. Seymour as his telegraphist-air-gunner (TAG), shot down a Dornier—the first enemy aircraft to be shot down by an Allied aircraft in the war.

That Dornier was shot down, but *Ark Royal* had been reported by the others, and the expected air attack developed in the afternoon. One Heinkel approached *Ark* and dropped one large 2000-lb bomb. The bomb, which took a long time to drop, looked to one observer 'as big as a pillar box'. To another, it was as 'big as my Austin Seven'—an opinion which, as the bomb fell nearer, he changed to 'as big as a double-decker bus'. After taking what 'seemed about ten years to strike' the bomb exploded thirty yards off *Ark Royal*'s port bow. The ship heeled over to starboard, a towering column of water crashed down on her flight deck, while a thick cloud of loosened soot flew upwards from her funnel. But the ship slowly righted herself and the only damage was to mess crockery!

The German pilot, Adolfe Francke, reported only a near-miss, but Dr Goebbel's propaganda machine embroidered his story and proclaimed that 'Der Flugzengtrager' was sunk. Francke was promoted to Oberleutenant, awarded the Iron Cross First Class, and commissioned to write a book for children on 'How I sank the *Ark Royal*'. The egregious Lord Haw Haw meanwhile asked every night, and indeed many times every night, 'Where is the *Ark Royal*? Britain ask your Admiralty, "Where is the *Ark Royal*?"' His broadcasts were much enjoyed by *Ark*'s ship's company who tirelessly chanted back 'We're here, we're here, we're here!' (However, such was the suggestive force of German propaganda that the Admiralty had eventually to take serious action through American and other neutral presses to show that *Ark Royal* was indeed still afloat.)

For the rest of 1939, the carriers were dispersed over the oceans of the world, searching for enemy surface vessels and engaged in various trade protection tasks. *Furious* had a very lucky escape indeed in December. With the battleship *Resolution* and the battle-cruiser *Repulse*, she was escorting an eastbound convoy of five large liners including the *Mauretania* and the *Aquitania*, carrying some 7500 men of the 1st Canadian Division. At 4 a.m. on 13 December, the westbound liner ss *Samaria* passed right through the convoy, carrying away *Furious*'s starboard side W/T masts. She then passed so close down *Aquitania*'s port side that she tore away five of her lifeboats. Finally she steamed very close down the starboard side of *Mauretania*. *Furious* and *Samaria*, both weighing 25,000 tons and both doing about 20 knots, very nearly collided head on, with two more giant ships piling into them from astern. This was all caused by *Samaria*'s outward routeing authority in Liverpool not being informed of the troop convoy's movements. What might have been the worst maritime disaster of the war, perhaps of all time, was only narrowly avoided.

The River Plate

While *Furious* was having her adventures in the North Atlantic, *Ark Royal* in Force K took part in the great search across the South Atlantic for the raiding German pocket battleship *Graf Spee*. *Ark Royal* was never within a thousand miles of *Graf Spee* but her aircraft searched over millions of square miles, rarely sighting anything but always having a powerful psychological effect upon the enemy, convincing him that discovery and danger might lurk over the next horizon. *Ark Royal* was part of the 'huge' fleet so ably conjured up by Winston Churchill and the BBC off the entrance to the River Plate, which was actually 'sighted' by *Graf Spee*'s gunnery officer and which eventually came to weigh so heavily on Captain Langsdorff's mind.

The first spotting operation of the war by a Fleet Air Arm aircraft (and, of course, the first since Flt Lt R.J. Rutland and Assistant Paymaster G.S. Trewin's Short seaplane at Jutland on 31 May 1916) was by the light cruiser *Ajax*'s Fairey Seafox seaplane during the battle of the River Plate. The Seafox, piloted by Lt E.D.G. Lewin, with Lt R.E.N. Kearney as his observer, was catapulted at 6.37 a.m. on the morning of 13 December. *Graf Spee* had already been sighted and had opened fire, with *Exeter, Ajax* and *Achilles* replying. Deafened by the blast from *Ajax*'s two after turrets which had been firing on a forward bearing, Lewin and Kearney were glad to get away. It was a fine clear morning, with visibility almost unlimited. *Exeter*'s two Walrus amphibians had been damaged by *Graf Spee*'s fire, and the whole task of spotting for Cdre Harwood's ships devolved upon Lewin who positioned his Seafox on *Ajax*'s disengaged bow, height 3000 feet, just below a layer of cloud. For all Lewin knew, *Graf Spee*'s aircraft might be launched at any moment and the cloud would give handy cover.

Exeter had received another two hits from *Graf Spee*'s 11-inch guns and had almost disappeared in a shroud of smoke and flame. Lewin and Kearney both thought she had gone but she emerged, still firing. But by 7.30 a.m. she could keep in action no longer and steered south and east to make repairs.

Ajax and *Achilles* in the meantime had been engaged by *Graf Spee*'s secondary 5.9-inch guns. The German gunnery, as usual, was excellent and *Ajax* was hit aft, losing both her after turrets. Lewin's Seafox, spotting for both *Ajax* and *Achilles*, saw *Graf Spee* apparently turning south to chase *Exeter*, but renewed fire from the two light cruisers made the enemy retreat to the north. *Graf Spee* herself had been hit and Lewin closed her to examine damage. The battleship replied with close-range anti-aircraft fire and Lewin retired again, after counting some thirty shell hits.

Shortly afterwards, Kearney saw *Graf Spee* launch several torpedoes at about five miles' range and warned Cdre Harwood who turned *Ajax* and *Achilles* towards *Graf Spee*. At 7.40 a.m. the Commodore steered east under cover of smoke and broke off the action, but continued to shadow the enemy when she eventually headed west. Lewin was ordered to find *Exeter* some eighteen miles to the south and signal her position, course and speed. After that, Lewin flew back to *Ajax*, landed safely alongside, in spite of the considerable sea running, and was hoisted inboard.

On 17 December, when *Graf Spee* was reported to have weighed her anchor and appeared to be putting to sea, Lewin's Seafox was launched again and took up station on *Ajax*'s starboard bow, ready for the action which seemed imminent. Just before sunset, Lewin saw two huge explosions fore and aft in *Graf Spee*, and for a moment he thought she was firing her main armament. But after a third explosion amidships Lewin signalled at 8.54 p.m. '*Graf Spee* has blown herself up.' 'It was now dark', Harwood wrote in his dispatches, 'and she was ablaze from end to end, flames reaching almost as high as the top of the control tower, a magnificent and most cheering sight.' *Graf Spee* had been scuttled and blown up on the orders of Captain Langsdorff who later committed suicide. Lewin received a DSC, and Kearney was mentioned in dispatches, the Fleet Air Arm's first honours of the war.

2

Operations in 1940

Norway

The Norway campaign in 1940 was a series of embarrassments for the Allies, of priceless intelligence information ignored or misinterpreted, of moves made wrongly or too late, and above all, of constant interference from afar with decisions of the men on the spot. Norway would have been an embarrassing possession, had the Allies won it. As it happened, she was lost in the most embarrassingly incompetent manner.

Furious had stayed in the Clyde for the first bitter winter of the war, until February, when she went to Plymouth for a refit. Her aircraft flew off to Abbotsinch, near Paisley; later, because they were grounded by snow and fog for weeks on end, they moved to an improvised airfield, occasionally used by Scottish Airways, at Campbeltown, on the Mull of Kintyre. *Ark Royal* meanwhile had stayed mostly at sea, on a variety of tasks including a search for the *Graf Spee*'s supply ship *Altmark*, of infamous memory, which was finally boarded, and her prisoners released by a party from the destroyer *Cossack*, on 15 February 1940. *Furious* returned to the Clyde on 28 March, by which time *Ark Royal* was in the Mediterranean, exercising with *Glorious*. *Furious* was therefore the only carrier in home waters when Germany invaded Norway.

Norway provided the Fleet Air Arm with its first real campaign test but, curiously, it had to press its services at first upon an apparently reluctant Home Fleet. No official orders came for *Furious* to raise steam and at noon on 7 April Captain Tom Troubridge eventually had to signal the Commander-in-Chief for permission to join him. Meanwhile a direct signal from the Admiralty arrived ordering *Furious* to join the fleet, which, due to the delays, she was not able to do until 8 a.m. on 10 April, having flown on the eighteen Swordfish of 816 and 818 Squadrons from Campbeltown on the way.

Furious sailed without her fighters, which was like sailing a battleship without her armour plating. According to the official history of the war at sea, this was 'because of the pressure from London to get all our forces to sea as quickly as possible'. But her Executive Officer, Cdr

Godfrey French, had another explanation in his recollections after the war: 'since the re-armament of *Furious* in 1938 it had never been possible to use the lower flying-off deck, and her maximum carrying and operating capacity of aircraft was reduced to eighteen Swordfish. There were no safety barriers, catapults, or any of the modern aids to take-off at this time. The carriage of aircraft on the flight deck was therefore out of the question. Therefore the question of embarking the fighters as well as the two squadrons of Swordfish never arose. The Commander-in-Chief had already decided that 801 and 804 Squadrons should remain at Hatston, partly for the protection of Scapa and partly to carry out operations in the south of Norway, for which the Skuas had the range but the Swordfish did not.'

From its first day, the Norwegian campaign began to explode old theories. It was the first time the Royal Navy had ever experienced a prolonged and determined attack from a skilful and well-provided enemy air force. The losses began at once when, on 9 April the battle-cruiser *Renown* was near-missed and the destroyer *Gurkha* sunk. Admiral Forbes, the Commander-in-Chief, who had relied on RAF support, found that air attack was already influencing his decisions, pre-empting the options open to him, and governing the disposal of his ships.

By the time *Furious* joined the fleet, the Skuas at Hatston had already carried out one brilliant but now almost forgotten feat of arms. The German cruiser *Konigsberg* had been badly damaged by Norwegian shore batteries at Bergen and was reported to be lying, immobilised, alongside the jetty. At 5.15 a.m. on 10 April, sixteen Skuas, seven from 800 Squadron led by Captain R.T. Partridge, RM, and nine from 803 Squadron, led by Lt W.P. Lucy, RN, each armed with a single 500-lb semi-armour piercing (SAP) bomb, took off from Hatston in the Orkneys. Bergen was some 300 miles away across the North Sea, a round-trip of 600 miles which was only just within the Skuas's operational range. They arrived over the Norwegian coast shortly after 7 a.m. within a few minutes of their ETA. The morning was fine, calm and cold. The water of the fjord lay like a polished silver mirror. The sun was just rising, like a great red orange, from behind the snow-covered mountains. Far below was a trace of haze above the town of Bergen and *Konigsberg*, looking like a thin slip of silver-grey metal beside the jetty.

The Skuas formed into line astern and climbed to 8000 feet for their attack. They had long practised their dives, and this was their moment. 'The ship was very clear and plain in my sights,' writes Partridge in his own account in *Operation Skua*, 'and the only opposition was one AA gun on the fo'c'sle manned by a very brave crew that contin-

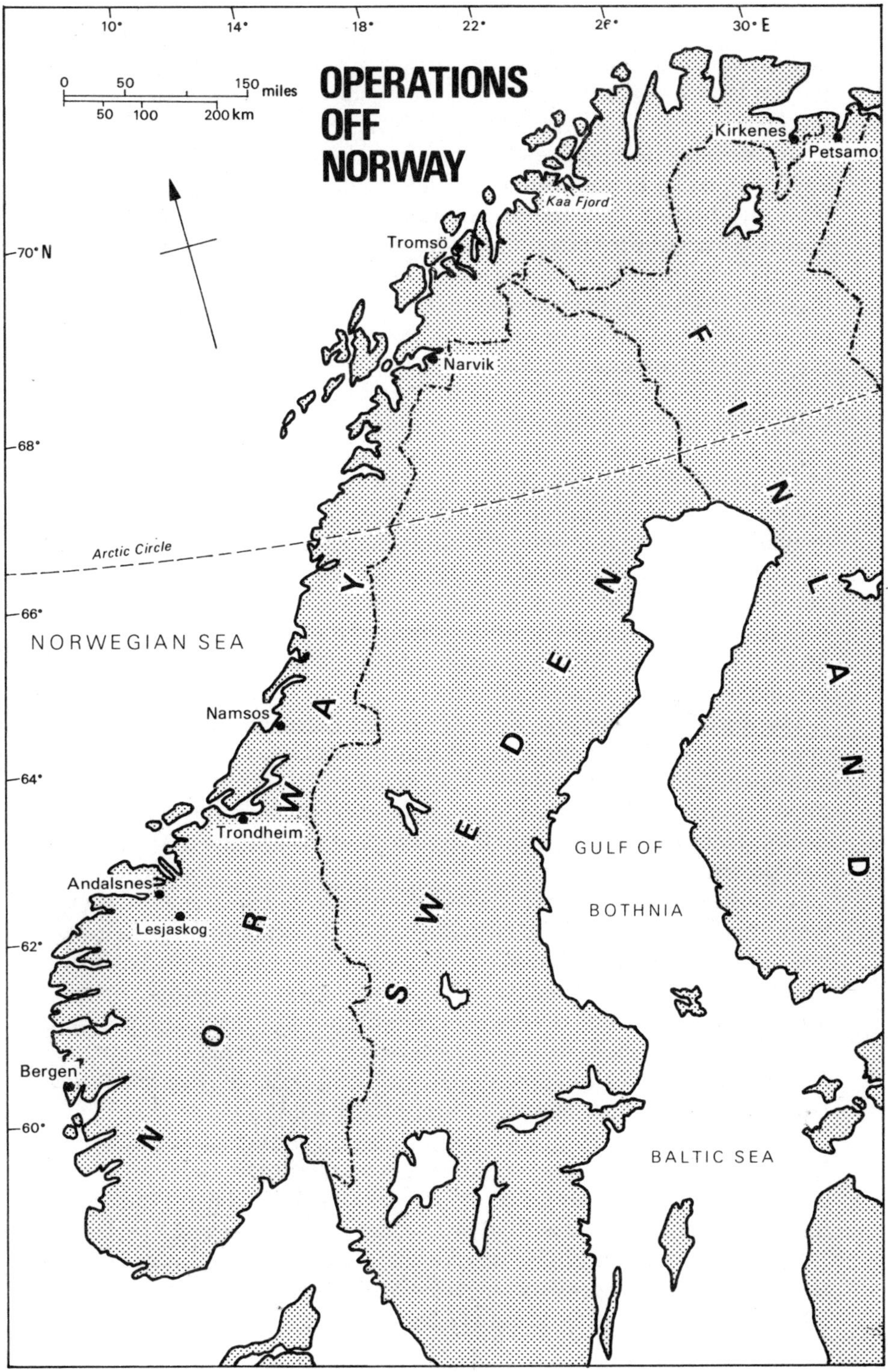
OPERATIONS OFF NORWAY
10°
14°
18°
22°
30° E
70° N
68°
66°
64°
62°
60°
0
50
150 miles
50
100
200 km
Kirkenes
Petsamo
Kaa Fjord
Tromsö
Narvik
Arctic Circle
NORWEGIAN SEA
Namsos
Trondheim
Andalsnes
Lesjaskog
Bergen
NORWAY
SWEDEN
FINLAND
GULF OF BOTHNIA
BALTIC SEA

ued firing throughout the whole attack. Down now to 4000 feet and still in that beautifully controlled dive that the Skua with its huge flaps could give. AA gun still firing and the tracer bullets were drifting up towards us like lazy golden raindrops going the wrong way. Now 2500 feet, no fear or apprehension, just complete and absolute concentration; mustn't drop too high and must watch going too low and blowing myself up with my own bomb blast. Very disturbed water round the ship, and water and oil seemed to be gushing out amidships. Still the fo'c'sle gun continued to fire and at 1800 feet I dropped my bombs and was away towards the sea at nought feet. My observer reported that we had had a near miss on the ship's port bow.'

Five bombs landed between thirty and forty yards of *Konigsberg*'s stern, rocking her violently and spraying her quarter deck and after superstructure with splinters. Five bombs landed on Skoltegrund Mole alongside the ship. Two exploded in the water in the narrow space between the ship's side and the Mole. Three were direct hits: one forward on 'A' turret, one amidships between the funnels, and the third on the port quarter. *Konigsberg* at once burst into flames along her entire length and a great column of smoke rose in the air. She began to list to starboard, while serious fires broke out below. Some fifty minutes later, watchers on shore heard another large explosion. *Konigsberg* broke in two and sank.

The Skuas meanwhile flew down the fjord, gunning targets of opportunity as they went, and re-formed over Lyso Island, at the fjord entrance. Three Skuas had been damaged and one lost: Lt B.J. Smeeton, and Mid F. Watkinson were missing. The remainder flew back to Hatston, landing on at 9.45 a.m. after a flight of some four and a half hours.

It had been an historic attack. *Konigsberg* was the first major warship ever to be sunk by enemy air attack. Yet, at home, the news was received in total silence, except for the BBC, who announced that night: 'The RAF have done it again!' Even today, the exploit is still almost totally unknown. However, the Skuas were prophets without honour only in their own country. Later in the war, the German *Kriegsmarine* and the Japanese Naval Staff both showed that they had fully digested the lessons of *Konigsberg*'s loss.

In the early hours of 11 April, *Furious* launched the first co-ordinated torpedo bomber strike in sea warfare. The German heavy cruiser *Hipper*, and four destroyers, were believed to be in Trondheim harbour and 816 Squadron, led by Lt-Cdr H. Gardiner, and 818, led by Lt-Cdr J. Fenton, flew off to deliver their attack at first light. In fact, *Hipper* had left the previous evening and only three destroyers were left. Tactically the sortie was a disappointment; because of the shallow water, none of the torpedoes (fitted with duplex pistols in their warheads, one of the very first occasions they were used operationally) hit any of their targets. However, all the Swordfish returned safely.

On the evening of the next day, nine Swordfish from each squadron took off to attack shipping at Narvik, each aircraft armed with four 250-lb and four 20-lb bombs. Flying conditions were appalling, with snow and sleet storms driven by winds of 30 to 40 knots, visibility sometimes coming down to less than a quarter of a mile, and a cloud ceiling of 1000 feet, often descending to 200 feet. *Furious* signalled the results to the Commander-in-Chief: 'Enemy seen in Narvik five *Maas* class destroyers and eleven merchant vessels. One destroyer alongside Iron Ore Jetty, three alongside pier and one at anchor. All attacked by 818 Squadron, two hit, one seriously. AA batteries on point and neck of Narvik Peninsula. Fire of ships accurate, batteries poor. Batteries seen halfway up hill behind Narvik. Photographs taken as ordered but visibility bad in Ofot Fjord. Verticals and obliques taken of Narvik. 816 Squadron ran into ceiling 200 feet with snowstorms so returned and just managed to land on before dark. One crashed over board on landing. All officers and men accounted for. Sub-Lt Roberts slightly wounded and Leading Airman Skeets bullets in leg'. Six Swordfish were damaged, two ditching in the sea.

On 13 April, *Furious* mounted a strike of ten Swordfish, all that were serviceable, to co-incide with the entrance of *Warspite* and nine destroyers into Narvik Fjord when they sank eight German destroyers by gunfire and/or torpedoes. This was, in a way, revenge for the first battle of Narvik, on 10 April, when Captain B.A.W. Warburton-Lee, of *Hardy*, lost his ship and his life but won a posthumous Victoria Cross. *Warspite*'s own Swordfish, fitted with floats, and piloted by PO F.C. Rice, with Lt-Cdr W.L.M. Brown as observer, was launched at 11.52 a.m. as *Warspite* was steaming up Ofot Fjord. With low clouds overhead and steep cliffs on either side it was 'like flying in a tunnel' but after sighting and reporting one German destroyer lurking in a small bay, they discovered a U-boat at anchor off Bjerkvik. Rice climbed to 300 feet and then dived to release two bombs, both of which hit, while his TAG Leading Airman M.G. Pacey raked the submarine's casing with machine-gun fire. The U-boat returned the fire and hit the Swordfish's tailplane before sinking. The U-boat, U-64, was the first to be sunk by a Fleet Air Arm aircraft. Meanwhile, two Swordfish were lost from *Furious*'s strike.

The 15 April was another busy day for *Furious*, best summed up in Captain Troubridge's signal to Flag Officer Narvik, which gives an excellent account of the range of tasks the aircraft undertook: 'Today's activities include photographic reconnaissance Narvik, machine fired on by AA guns and automatics. A/S patrol Vaag Fjord during landing of troops. Unsuccessful A/S search for reported submarine. Attack by nine Swordfish on nine enemy aircraft suspected Heinkel bombers but proved to be Junkers 87C troop carriers, apparently immobilised on Frozen Lake. Two put out of action, others machine-gunned, enemy returned fire, one Swordfish force landed (crew safe) remaining machines returned. Mid Bland, RNVR, slightly wounded.'

The last aircraft from this strike landed on two hours after sunset, although the pilot had never deck-landed by night before. The crew of the ditched Swordfish was picked up by the destroyer *Zulu*.

Throughout all these operations, the fleet had no fighter cover and *Furious* thus had to operate some miles offshore, adding to the navigational difficulties of her aircraft. *Furious*'s Swordfish, like those on later carriers, were the Army's almost sole air support, for bombing and reconnaissance duties. The Swordfish flew in terrible conditions, of snow, sleet and rain, with normally low visibility and cloud ceilings, along narrow fjords and between steep mountain sides. In low visibility it was often difficult to distinguish between cloud and snow-covered rock. The shape of the fjords swept sudden wind squalls across the water surface, throwing an unwary aircraft off course.

To the dangers of the weather and the sea were added the violence of the enemy. The Swordfish suffered constantly from the attentions of German fighters and frequently found, to their amazement, that their one advantage was their slowness. Again and again the much faster Me109s overshot in their attacking runs. Nor was the action over when they returned to their carrier, for as the days passed, the fleet was subjected to the full ferocity of the Luftwaffe's attacks. The losses mounted. The destroyer *Eclipse* was bombed and badly damaged on 11 April, and on 18 April the cruiser *Suffolk* struggled home to Scapa with her quarterdeck awash after seven hours of bombing. *Furious* was normally lucky. As Lt-Cdr Jenkins, her navigating officer, wrote in *Days of a Dogsbody*, a few minutes before enemy aircraft were due, 'up would come a convenient snow-blizzard in which the ship would be completely lost to sight for the essential half-hour or so before their shortage of petrol forced them to return. Whereupon our friendly blizzard would roll away, in nice time to save me from too much anxiety as regards running ashore in those relatively narrow waters.'

But, on that 18 April, *Furious*'s luck almost ran out. A solitary Heinkel dropped two large bombs, one of which exploded close on the port side, shaking *Furious*'s port turbine blading out of alignment and reducing her speed to 20 knots. By 25 April, when *Furious* withdrew, her aircraft, as Captain Troubridge signalled, had 'flown 23,870 miles, dropped eighteen torpedoes, 15½ tons of bombs, lost fifty per cent of planes and had seventeen hit by enemy fire, taken 295 photographs. Casualties—three killed, seven wounded, two missing feared dead.'

The day before *Furious* left, *Ark Royal* and *Glorious* arrived from the Mediterranean. *Ark Royal* had 800 and 803 Squadrons, each with twelve Skuas, 801 Squadron with six Skuas and six Rocs, and two Swordfish squadrons, 810 and 820, each with twelve aircraft. *Glorious* had a squadron of twelve Sea Gladiators, 802, and one of six Sea Gladiators, 804, six Skuas of 803, and twelve Swordfish of 823 Squadron. She also carried the Gladiators of 263 Squadron RAF which were flown off to operate from the frozen Lake Leskaskog; so it seemed that, at last, fleet and army would have both carrier-borne and shore-based fighter cover. However, all the Gladiators of 263 had either been destroyed or were out of action within forty-eight hours, so once again the brunt fell on the carriers.

Operating about 100 miles offshore, because of their own inadequate fighter cover, *Ark Royal* and *Glorious* maintained fighter patrols over convoys and coastal areas, besides keeping a Combat Air Patrol (CAP) over Namsos and Andalsnes, while Swordfish raided Trondheim, bombed Vaernes airfield and the frozen lake of Jonsvatnet some five miles from Trondheim. Once again, there was considerable German fighter opposition and the Skuas proved much too slow in combat against the Messerschmitt Me109s. On 25 April alone, five aircraft were lost to enemy action and another five operationally—lost largely because of the distance the carriers had to stay offshore. Several other aircraft were damaged by fighters or shore batteries.

Lt R.C. Hay, Royal Marines, was on his first patrol in a Skua of 801, led by Lt Bill Squires, when 'we ran into a Heinkel IIIK bomber. Bill attacked from astern and the bomber dived to sea level. They exchanged fire and when Bill pulled upwards to break off the attack, his aircraft was struck in the belly and crashed in the sea without survivors. This was the first lesson in airfighting with a vengeance—never break away upwards. I therefore sat on the tail of the bomber and fired short bursts until it crashed into the sea.'

The Roc, with its ball-turret and no front-firing gun, was at an inherent disadvantage in dog-fighting as Hay found out one day patrolling above the fleet. 'Soon we spotted a Ju88 reporting the fleet's position. Diving down at maximum speed we got to within 200 yards of the bomber—but I had no front guns so I attempted to slew my aircraft to enable the turret to bear. Shouting "Fire" to the gunner, I heard one bang and silence. All the guns had jammed. The German fled with a startled puff of brown smoke from his diesel engines. . . .'

On 26 April Partridge of *Konigsberg* fame shot down a Heinkel over Gudbrandsal but his Skua was also hit and he had to force-land on a frozen lake. He and his observer, Lt R.S. Bostock, burned their confidential papers and signals, set fire to their Skua and then struggled through deep snow to a tiny hut they could see on a hillside not far away. They had lit a fire and settled down as comfortably as they could for the night, when they heard footsteps, voices and a whistle. Standing outside they discovered the three-man crew of the Heinkel they had shot down. It was a delicate situation, not covered in books of etiquette, but although one of the Germans was armed, Partridge convinced them they were his prisoners and all five spent the night in the hut. A Norwegian patrol next morning very nearly shot them all as Germans, but later Partridge and Bostock returned to the ship.

Skuas from *Ark Royal* and Gladiators from *Glorious* covered the evacuations of Namsos and Andalsnes. *Ark Royal* was near-missed twice more on the same day, 1 May, but her incredible luck still held. The nearest bomb, a 500-pounder, fell only ten yards off the starboard side but failed to explode. *Ark Royal* returned to Scapa on 3 May, where several missing pilots rejoined, amongst them Mid L.H. Gallagher. He had failed to find *Ark* after a raid so returned and landed amongst the wreckage of 263 Squadron Gladiators on Lake Lesjaskog. Here he topped up his Skua with petrol but could not find a cartridge for the impulse starter of his engine. Borrowing a rifle, he served for the next three days with the troops ashore. On the fourth day, he found a wrecked Skua, with an impulse starter cartridge on board which he took back to Lake Lesjaskog and his engine fired successfully. He took off safely and flew back alone 350 miles to Hatston. When he was asked how he had navigated his aircraft, he is supposed to have replied, with one of the most monumental lines ever shot by any pilot, 'Well, actually I borrowed an atlas from a Norwegian kid and, er, sort of used that'.

Later in the month *Ark Royal* moved up to Narvik, where since 18 April a squadron of six Walrus amphibians of 701 Squadron, based at Harstad under Cdr R.S.D. Armour had been carrying out A/S patrols, providing convoy cover, investigating intelligence reports and making over 200 communication flights, ferrying British and French officers. These Walruses made one final bombing attack on a concentration of German troops before embarking in *Ark Royal* on 8 May.

Off Narvik, *Ark Royal* flew Swordfish photo-reconnaissance flights, as well as providing fighter cover, and bombing support for the troops ashore. On 8 May, one Swordfish piloted by Lt H. de G. Hunter was ambushed by three German aircraft. While his TAG Leading Airman Bennett fought them off, the observer Lt A.W.N. Dayrell signalled to *Ark* 'From Swordfish 4F "Delayed by three Heinkels"'.

On the 14 of the month, *Ark Royal* lost one of her best pilots, Bill

Lucy, also of *Konigsberg* fame, who was shot down while following a damaged Heinkel close to the sea. He and his observer Lt M.C.E. Hanson were both killed. After covering the Allied assault and capture of Bjerkvik, *Ark Royal* returned to Scapa again on 21 May, at a time when the force of German air attacks was increasing.

The effect of the constant air attacks was described by AB Joseph Lawrenson, one of the guns' crews of the anti-aircraft cruiser *Curlew* in 'The Cruiser *Curlew*': 'There were dive-bombers which threatened to take our masthead with them in every screaming nerve-racking dive. There were the lazy, seemingly indifferent, high-altitude bombers whose pilots kept just out of our gun range, scattering their bombs with the indifference of a child tossing posies; and the annoying bumble-bee type of plane which sat still in the sky and watched us from a distance.'

After *Ark Royal*'s departure, several warships, transports or store-ships were sunk or damaged. On 26 May, 'a beautiful Sunday afternoon', *Curlew* herself whose radar had been of priceless help to the ships offshore was hit and sunk. 'The ship seemed surprised. Engines stilled; a deadly silence fell; we realised that it was all over. A horrible gurgling and a groaning began; a Maltese cook ran across the upper deck trying to push back his hand that had been neatly sliced off; a young boy of fifteen manned the now vacant machine-gun position and blindly shot the sky; a body draped over the guard-rail.'

Furious returned on 21 May to fly off a reconstituted 263 Squadron of Gladiators to Bardufoss, and *Glorious* delivered Hurricanes of 41 Squadron a few days later. Narvik was taken on 28 May with the object of destroying the railway and the iron ore loading plant. This was achieved but the campaign was already lost for the Allies. The decision was made to abandon Norway and on 2 June *Ark Royal* and *Glorious* arrived off the coast again to cover the final evacuation.

The RAF crews of 263 and 41 Squadrons had to decide whether to destroy their aircraft ashore or fly them off to a carrier. To their great credit they chose to land on *Glorious*, although none of them had ever landed on a deck before. In so doing, those RAF pilots did a better turn for their hosts than they knew. They showed, once and for all, that high-performance aircraft such as Hurricanes *could* land on flight decks successfully.

There were other lessons to be drawn from the Norwegian affair. *Furious* had discovered the urgent need for some form of air operations' plot, to correlate and present air action information. *Ark Royal*, too, made some of the first experiments in fighter control, albeit at second-hand, using the radar set in the cruiser *Sheffield*. From these early efforts sprang the science of fighter-direction which was to be so well developed by the time British carriers operated in the Pacific. But the Fleet Air Arm's main successes in Norway were of much wider significance. The RAF in Norway had not been able to cover landings or evacuations, nor attack enemy shipping, nor co-operate with the Army as pre-war doctrine claimed it could. The Fleet Air Arm had done all these things. To the Navy as a whole, the determined enemy air opposition over Norway demonstrated that it was not possible to maintain the Army on an overseas expedition, or even to operate ships for any length of time off an enemy coast, without proper and sustained air cover.

Loss of *Glorious*

There was one last bitter lesson from the Norwegian campaign: that against capital ships an aircraft carrier not operating her aircraft is as helpless as a picket boat. On 8 June, *Glorious* was detached from the fleet, with two destroyer escorts *Ardent* and *Acasta*, to make her own way home independently, because she was short of fuel. At about 4 p.m. that afternoon her funnel smoke was sighted to the north from the German battle-cruisers *Scharnhorst* and *Gneisenau*, who were then steaming northwards, hoping to intercept Allied convoys leaving Narvik.

Glorious was not flying reconnaissance patrols, nor maintaining fighter cover overhead. Possibly her flight deck and hangar were congested by the extra RAF fighters which had flown on earlier that day. Or possibly her commanding officer, Captain G. D'Oyly Hughes (a very distinguished submariner, who had been Nasmith VC's First Lieutenant in E-11 in the Dardanelles in 1915), believed that their best way home was the straightest, without having to turn in and out of wind to operate aircraft. But whatever the reason, *Glorious* was caught completely unaware.

Scharnhorst opened fire at about 4.30 p.m., at a range of 28,000 yards, and as usual, the initial German gunnery ranging was excellent. *Glorious* was soon hit and damage to her hangar prevented the frantic efforts being made to arm and range her Swordfish. Marine Ronald Healiss, one of a 4.7-inch gun's crew, described *Glorious*'s last moments in *Adventure Glorious*: 'The flight deck above us was wreathed in smoke, then tongues of flame, then the staccato sounds of fire rose to a great roar, and a red wall like a living furnace rose from the hangar-well . . . I looked up the cliff face of the wall of steel forward of the bridge and saw the steel shattering. Plummets of smoke filled the air as I watched in horror. They'd hit the bridge. . . . It was that instant another salvo hit, and the whole side of *Glorious* seemed to cave in, leaving a choking cloud of smoke and a thunderous roar that echoed away to the darkening sky. The sea, so calm before the action, was now churned up and flecked with grey.'

Ardent and *Acasta* both tried with great bravery to save *Glorious*, laying a protective smoke screen and then advancing through it to attack with torpedoes, but it was all in vain. *Glorious*, stopped and on fire, turned over to starboard and sank at about 5.20 p.m. Eight minutes later, *Ardent* was also sunk, having fired all her torpedoes. *Acasta* lasted until 6.08 p.m. when she too sank – but one of her torpedoes hit *Scharnhorst* abreast her after gun turret and badly damaged her.

The tragedy of *Glorious* did not end with the loss of the ship. The two German battle-cruisers steamed past the scene of the action, without pausing to pick up survivors. *Glorious* had made an enemy sighting report at about 4.15 p.m., but no ship or station received it. At 5.20 p.m. the cruiser *Devonshire*, some 100 miles to the west, heard a ghostly whisper out of the ether: 'Vice Admiral from *Glorious*. My 1615 two PB. Time of origin 1640.' The 'PB' obviously referred to 'pocket battleships', but the signal was deemed to be corrupt, and was not understood. *Devonshire*, taking the King of Norway to England, rightly preserved radio silence. No signal of any kind was received from either of the two destroyers and so the first that the Admiralty knew of the disaster was a German broadcast at 3 p.m. on 9 June.

Many men got away from the sinking ships on Carley floats, but because of the bitter cold and the heavy swell running (which capsized all *Acasta*'s boats) men soon began to die of cold, exhaustion and exposure. One float with twenty-two officers and men was reduced to four by the morning of the 9th. Aircraft from *Ark Royal* were seen by the survivors, but did not sight them. It was not until 12.30 a.m. on 11 June that the small Norwegian vessel *Borgund* picked up three officers and thirty-eight ratings of *Glorious* and one rating from *Acasta*, landing them at the Faroes on the 14th. One officer from *Glorious*, and four ratings, with another two from *Ardent*, were picked up by a German seaplane and became prisoners of war. These were the only survivors. Ninety-four officers and 1380 ratings from *Glorious, Ardent* and *Acasta* were lost, with another forty-one RAF personnel, making a total loss of 1515 lives.

The damage to *Scharnhorst* caused both ships to abandon their sortie and make for Trondheim where, on 13 June, *Ark Royal*'s Skuas slung a final Parthian shot at them. It was not a very effective shot, indeed the Skua attack on *Scharnhorst* turned out to be one of the Fleet Air Arm's saddest operations. There had been forebodings about it the night before. The forced gaiety of a party in *Ark*'s wardroom had soon broken up. Even Lt G.E.D. Finch-Noyes, one of the ship's most celebrated characters and star of the ship's concert parties, confided to a friend that he had misgivings about the day to come.

His feelings were fully justified. Fifteen Skuas of 800 and 801 Squadrons, each armed with a 500-lb bomb, took off just after midnight. There was not a shred of cloud cover and they had some 100 miles to fly after crossing the coast, where of course they were seen and reported. RAF Beauforts and Blenheims were expected to make synchronised bombing attacks, and to give cover, but there was no sign of them. There were plenty of German fighters and the flak barrage was fiercer than anything the pilots had experienced before. The Skuas dived from 11,500 feet for their attacks, some of them coming down almost to sea-level before trying to get away. Casson of 801 attacked from 3000 feet, Partridge of 800 from 7000 feet. One bomb hit *Scharnhorst* but failed to explode. Eight Skuas were lost. Casson, and Partridge, badly burned, were shot down and became prisoners of war. Amongst the casualties were Finch-Noyes and the resourceful Mid Gallagher.

Captain Troubridge's tribute (from his report of proceedings) to his own aircrews in *Furious* can be taken to apply to all the Fleet Air Arm crews who flew over Norway: 'It is difficult to speak without emotion of the pluck and endurance of the young officers and men, some of them midshipmen, who flew their aircraft to such good effect,' he wrote in his report. 'Once they had undergone their baptism of fire their morale and spirit rose as each obstacle was in turn successfully surmounted. All were firing their first shot in action, whether torpedo, bomb or machine-gun; many made their first night landing on 11 April; and, undeterred by the loss of several of their shipmates, their honour and courage remained throughout as dazzling as the snow-covered mountains over which they so triumphantly flew.'

The Battle of Britain

For a few weeks in the late summer of 1940 the fate of the country, and of Western Europe, lay in the hands of a few hundred fighter pilots. The Fleet Air Arm played a small but valuable part in the Battle of Britain. 804 Squadron, of Sea Gladiators, at Wick, and 808 Squadron, of Fairey Fulmars, at Castletown, came under the operational control of 13 Group RAF, with its headquarters at Newcastle-on-Tyne. But for some naval pilots, their involvement was much closer. They actually flew with the RAF in regular RAF Squadrons.

On 6 June, just after Dunkirk, forty-five naval pilots under training, including seven ex-RAFVR pilots with the Fleet Air Arm, were transferred temporarily to the RAF to complete their training. They were sent to operational training schools in order to convert to flying Spitfires and Hurricanes. Thirty more naval pilots transferred by the end of June, making a total of sixty-eight (the seven ex-RAFVR had rejoined the RAF). Ten naval pilots were recalled by the Navy for service in the Mediterranean in July 1940, so fifty-eight naval pilots actually took part in the Battle of Britain.

They wore naval uniform but for operations and discipline they were fully integrated into the RAF squadrons, and as casualties occurred, became section leaders, with RAF pilots under them. They served with some famous RAF units, 19 Squadron at Duxford, 213 Squadron at Tangmere, 46 Squadron at Stapleford and others. Three of them, Sub-Lt R.E. Gardner, Sub-Lt R.C. Cork and Mid P.J. Patterson served with 242 Squadron under Sqn Ldr Douglas Bader at Coltishall, north of Norwich.

They began with convoy patrols off the east coast and Gardner shot down a Ju88 on 16 July, after chasing it fifty miles out to sea. The Squadron was first fully engaged in the Battle of Britain on 30 August, when Cork shot down his first German aircraft. Cork, who later became a notable fighter pilot and one of the Fleet Air Arm's best known characters, flew in several engagements with Bader, including what Bader called 'the finest shambles I have ever been in' on 15 September, at the peak of the battle. Cork's combat report published in the Ministry of Information's *Fleet Air Arm* of 1943 still recaptures the urgency and the excitement of those sunlit desperate encounters, high above southern England. 'Whilst flying as Red 2 in the leading section of the squadron we sighted the enemy to the south and well above us. We climbed as fast as possible to the attack, but on the way were attacked from above and behind by a number of Me109s. The order was given on R/T to break formation, so I broke sharply away with a Me on my tail. I was now in a dive and suddenly flew through the second squadron in the Wing formation and lost the enemy machine; at the same time I saw a Do17 on my starboard, flying NW. I dived 6000 feet to attack and fired a long burst at the port engine, which started to smoke. I attacked again on the beam—large pieces of enemy machine flew off and his starboard wing burst into flames near the wing tip. He dived straight into the cloud, heading towards a clear patch, so I waited until he came into the open and fired another burst in a head-on attack and the machine dived into the ground.

'I climbed up 1000 feet and was attacked by two yellow-nosed Me109s from above, so I did a steep turn to left and managed to get on the tail of one, fired a very short burst, and then ran out of ammunition. No damage was seen on enemy machine, but as I was being attacked from behind by a second fighter I went into a vertical dive down to 2000 feet and returned to base. No damage to my own machine.'

Another pilot, flying from a squadron near Dover, had a most eventful first combat flight also reported in *Fleet Air Arm*: 'For the first time the Hun appeared in large numbers and in broad daylight. We (three of us) were directed over the Channel to meet a reported fifty plus of aircraft. We climbed rapidly to 20,000 feet, and then I saw them in the bright blue of the sky—three solitary white trails coming out of a high layer of cloud, and growing steadily longer. They were some 4000 feet above us. We gave our Hurricanes everything they had and climbed up towards them. Instead of attacking us they climbed rapidly and went into a defensive circle, making no attempt to attack. We reached them at last at 32,000 feet, our Hurricanes on the point of stall and with no manœuvrability at all. Still they did not attempt to attack us, although they had manœuvrability and about twelve times our number. We separated and I flew inside their circle, in the opposite direction to them, and waited until one flew into my sights. I fired and the recoil immediately stalled me and I started to spin. I got out in about 2000 feet, and looking about me found there wasn't an aircraft in the sky; in thirty seconds some forty aircraft had completely disappeared and I had the sky to myself.

'I think after the excitement and the exertion at that height I was a bit dazy, anyway I was rudely awakened by long flickering streams of tracer which seemed uncommonly near my head. Then he shot by in a steep dive, a lean black little blighter—my first 109. I dived as hard as I could make my Hurricane, at any moment expecting a wing to break off, and still the 109 drew away. Suddenly, near the French coast, he started to level off, and, quite satisfied apparently that he was safe, headed over the town of Calais. I got my sight on him, pressed, saw flickering tendrils reach out from my wings to his tail, saw a large piece come adrift and slip by me. Smoke started to trail, then crumpled, and I seemed to do a complete somersault. Flak had opened up at me, despite their own machine. I shot off home as fast as I could, having gained my first probable.'

By December 1940, when the naval pilots returned to the Navy, eighteen had been killed, including Mid Patterson and Sub-Lt F. Dawson Paul, RNVR, a gifted pilot who served with 64 Squadron RAF and shot down five German aircraft in July alone.

Taranto

'Laurels grow in the Bay of Biscay,' wrote Nelson, 'I hope a bed of them may be found in the Mediterranean.' He might have been writing of the last six months of 1940 when, after the entry of Italy into the war, the Mediterranean saw an amazing flowering of naval air power. Despite the collapse of France, on whose Mediterranean fleet so many Allied hopes and plans had been pinned, Admiral Cunningham, the Commander-in-Chief—a destroyer man born and bred, and a late convert to naval air power—had for a very brief period the rare luxury for a British admiral of a balanced fleet of capital ships and at least one and sometimes two aircraft carriers. Although his was always the weaker of the two fleets, by the end of the year Cunningham had won a startling psychological advantage over his Italian opponents.

The carrier *Eagle* came through the Suez Canal to Alexandria in May 1940, with nine Swordfish and three Sea Gladiators of 813 Squadron and nine Swordfish of 824. *Ark Royal* later arrived at Gibraltar and, on 28 June 1940, formed part of Force H—a group of capital ships and a carrier, with destroyer escort, later known as a task force—formed to fill the vacuum in the western Mediterranean caused by the disarming of the French fleet.

Ark herself played a part in that disarming. She then had two fighter squadrons, 800 and 803, with twenty-four Skuas, and three Swordfish squadrons, 810, 818 and 820, with thirty aircraft. On 3 July, *Ark*'s Swordfish spotted for the bombardment of the French ships at Mers-el-Kebir, and laid mines at the harbour entrance. When the modern, fast French battleship *Strasbourg* escaped, *Ark Royal* mounted two Swordfish strikes, both of six aircraft, the first with bombs, the second with torpedoes. Neither succeeded, and two Swordfish were lost, but this was the first time aircraft had attacked a capital ship with torpedoes in the open sea.

Meanwhile, *Hermes* had been shadowing the new battleship *Richelieu* who eventually took refuge at Dakar, in French West Africa. Early on 8 July, six Swordfish of 814 Squadron flew into Dakar harbour and in the face of strong AA fire obtained one torpedo hit on *Richelieu* aft, flooding three compartments and distorting one propeller. She was not repaired for over a year, although she could have got under way in an emergency.

Richelieu was still at Dakar, with her main armament still intact in September and her accurate gunfire was one of the factors which contributed to the failure of that abortive expedition (others being the loquacity of the Free French in London, and a vastly over-optimistic estimate of the degree of support de Gaulle commanded in French West Africa). The operation at Dakar, although militarily and politically misconceived, was the first occasion when a carrier, *Ark Royal*, provided the whole of the air support for an amphibious landing. Some valuable lessons were learned, the chief being that one carrier alone simply could not, as early as 1940, provide all the reconnaissance, spot-

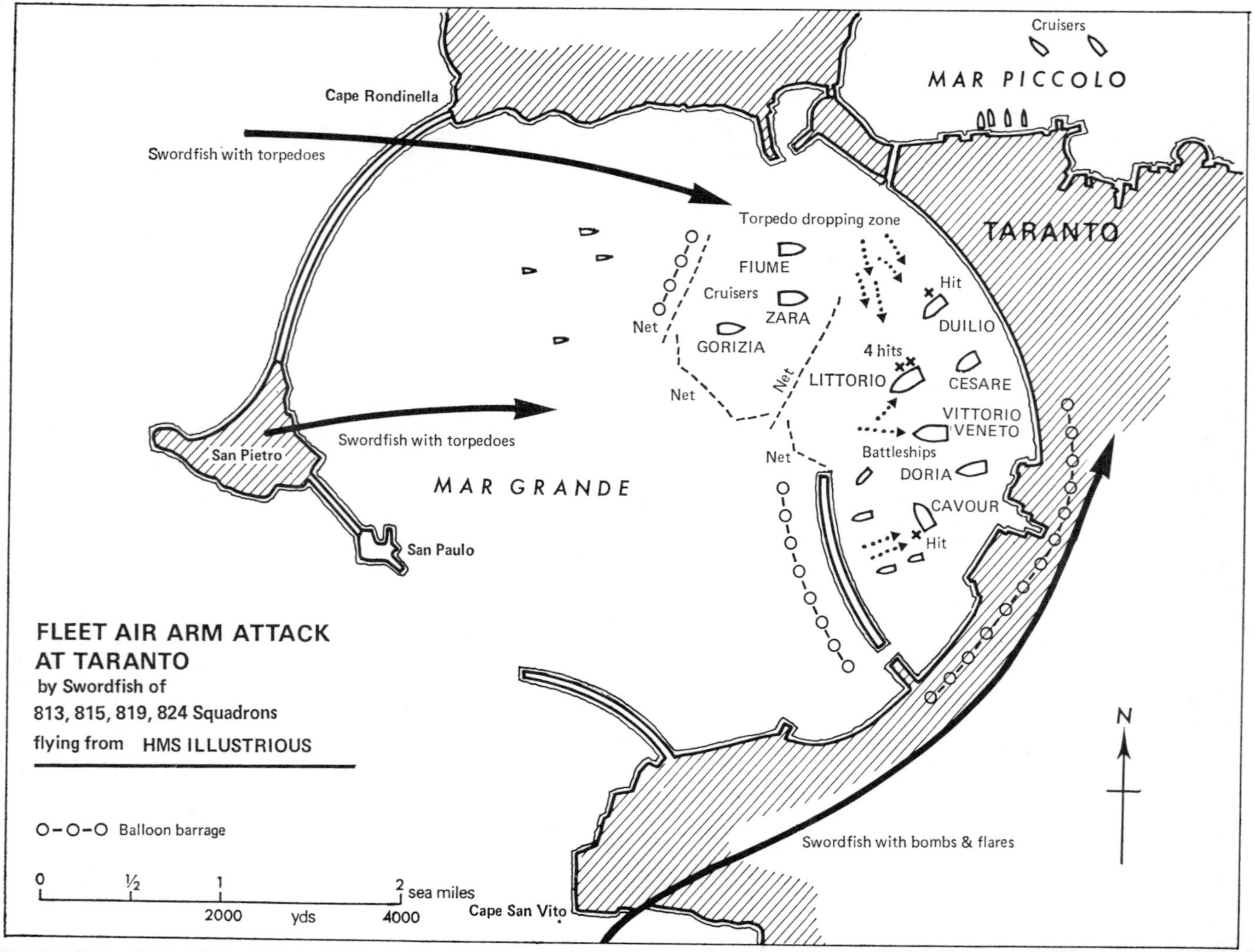
FLEET AIR ARM ATTACK
AT TARANTO
by Swordfish of
813, 815, 819, 824 Squadrons
flying from HMS ILLUSTRIOUS
O–O–O Balloon barrage
0
½
1
2 sea miles
2000
yds
4000
Cruisers
MAR PICCOLO
TARANTO
Cape Rondinella
Swordfish with torpedoes
Torpedo dropping zone
FIUME
Cruisers
ZARA
GORIZIA
Net
Net
Net
Net
Hit
DUILIO
4 hits
LITTORIO
CESARE
VITTORIO VENETO
Battleships
DORIA
CAVOUR
Hit
San Pietro
Swordfish with torpedoes
MAR GRANDE
San Paulo
Swordfish with bombs & flares
N
Cape San Vito

ting, fighter CAPS and torpedo attacks such an operation required. Swordfish, as those who knew them would confirm, were very slow and vulnerable against alerted targets in daylight. Once again, the ponderous Skuas were easily outperformed by their opponents, this time Vichy French Morane Saulnier 406 and Curtiss Hawk 75 fighters.

The first real brush with the main Italian battle fleet occurred off the south coast of Calabria on 9 July 1940, when the Mediterranean fleet was at sea to cover the passage of two convoys from Malta to Alexandria. Shortly after 3 p.m., Captain Rory O'Conor in the cruiser *Neptune* had the privilege of being the first Royal Navy captain in the Mediterranean since the Napoleonic wars to signal 'Enemy battle fleet in sight'. It was an historic day for *Eagle* too, for she became the first aircraft carrier in naval history to work with a fleet in battle. She launched a succession of Swordfish sorties against the Italian fleet, none of which scored, at the same time spotting for the battleships' guns. After *Warspite* got one hit at 26,000 yards on the battleship *Cesare*, the Italian fleet turned away. It was an inconclusive and somewhat disappointing engagement, but it had already begun to establish a morale advantage over the Italians. On the debit side, the Italian high-level bombing had proved to be disquietingly accurate.

The Italians were constantly kept off balance by the sheer unexpectedness of attacks. Naval aircraft were restless, irrepressible opponents, liable to strike literally out of the blue. One such raid in August, by three of *Eagle*'s Swordfish from 824 Squadron, can only be described as brilliant but cheeky.

Followed by an ancient Vickers Victoria carrying their maintenance ratings, wheel chocks, tools, spare parts and torpedo equipment, the three Swordfish were sent ashore to operate for a period from Ma'aten Bagush, the RAF HQ in the western desert. Led by Captain Oliver Patch, Royal Marines, the three Swordfish took off at 7 a.m. on 22 August and, after fuelling at Sidi Barrani, flew some fifty miles out to sea, to avoid Italian fighters. At 12.30 p.m. they turned towards the land and flew at thirty feet into the harbour of Bomba, near Gazala, on the coast of Cyrenaica, where reconnaissance flights and intelligence intercepts had reported ship targets.

Patch himself torpedoed and sank the Italian submarine *Iride*. His wingman Lt (A) N.A.F. Cheesman was about to attack a second submarine when his observer Sub-Lt F. Stovin-Bradford noticed they were over shoal water and, just in time, prevented Cheesman from dropping his torpedo in the sand. Cheesman coolly flew on another 350 yards in the face of quite fierce AA fire before completing the attack. Cheesman and Patch's other wingman, Lt J.W.G. Wellham, between them sank a destroyer and a submarine depot ship, and badly damaged the second submarine. However, Wellham's Swordfish was damaged. It reached Sidi Barrani but Wellham had to hitch a lift with Patch from there on. Later, it transpired that *Iride* had been just about to sail with frogmen of the Italian 10th Light Flotilla to attack the fleet at Alexandria.

At the end of the month, *Illustrious* arrived in the Mediterranean. She was the first of a new, fast, 23,000-ton class of carrier, with three propellor shafts and an armoured flight deck. Her squadrons were 815 and 819 of nine Swordfish each, who had already operated together from Detling in Kent whilst helping the RAF during the Dunkirk evacuation. Her fighter squadron was 806, of fifteen Fairey Fulmars, the new eight-gun 270 knot interceptors (still, however, two-seaters).

The Fulmar was slow and not particularly manœuvrable for a fighter and inspired one or two derisive verses, such as

Any old iron, any old iron,
Any, any, any old iron;
Talk about a treat
Chasing round the Fleet
Any ole Eyetie or Hun you meet!

Weighs six ton,
No rear gun
Damn all to rely on!

You know what you can do
With your Fulmar Two;
Old iron, old iron!

However, despite their short-comings, Fulmars made 112 kills in two years.

To aircrew who had served in the old pre-war carriers, *Illustrious* herself was an eye-opener. As Lamb, who joined her in 815, said 'everything in her was so much bigger and better. . . . Comparing her with *Courageous* was tantamount to comparing her with Noah's Ark.' Amongst other refinements, *Illustrious* had a wire safety barrier, to protect the forward deck park whilst aircraft were landing on aft. By raising and lowering the barrier, the landing intervals between successive aircraft could, with practice, be cut to a few seconds.

Used with great intelligence and aggressiveness, the fleet's aircraft struck at the enemy from Rhodes in the west to Sfax in Tripolitania in the east, ranging over the harbours and airfields of the Dodecanese islands, strafing and bombing targets at Scarpanto and Stampalia. The Fulmars proved a mite slow for fighter work, but they were quite capable of dispatching Italian flying boat shadowers and high-level bombers, and they put up such an effective umbrella over the fleet that for some months no ship was lost to enemy air attack. The climax and crown of what now seems in retrospect a brief halcyon period of carrier power was a famous feat of arms—the Fleet Air Arm's first and greatest battle honour, the night strike on the Italian battle fleet at Taranto.

Plans for a strike on the Italian battle fleet at Taranto had been considered as early as 1935, during the Abyssinian crisis. By the autumn of 1940, the reluctance of the Italians to put to sea made such a strike even more attractive; if the Italians would not come out, they must be hit where they hid. But first, certain conditions had to be fulfilled. There had to be what the Commander-in-Chief called 'good and timely photographic reconnaissance of the harbour'. Flying boats had proved too slow and vulnerable for this task, which had to await the arrival at Malta of Glenn Martin Marylands of No 431 Flight, RAF. Further, the striking Swordfish would have to be fitted with long-range fuel tanks, so that the operating carriers would not need to approach too close to the enemy coast. Such tanks were not available until *Illustrious*'s Swordfish joined the fleet in September 1940. Lastly, but most importantly, the pilots and observers had to be adequately trained in night flying.

By October 1940 these requirements had all been met and a strike by Swordfish from *Eagle* and *Illustrious* was planned for the 21st—Trafalgar Day. But it had to be cancelled, because of a hangar fire in *Illustrious* a few days before the target date, and because of defects in *Eagle*'s fuel system which had been thoroughly shaken up by several bomb near-misses—an indication of how uncomfortably close the much-despised and derided Italian high-level bombing had actually been. *Illustrious* herself suffered from water in the aircraft fuel system and lost three Swordfish which had to force-land in the sea from this cause. In the event, *Eagle* did not take part, but five of her Swordfish and twelve aircrew, six pilots and six observers of her 813 and 824 Squadrons transferred to *Illustrious* for the strike.

Photo-reconnaissance flights were flown on 10 November 1940. The photographs, when they had been collected from Hal Far, in Malta, and flown out to *Illustrious* at sea, showed that the Italian fleet was at home, with six battleships in the harbour. At 6 p.m. on 11 November *Illustrious*, flying the flag of Rear Admiral Aircraft Carriers, Rear Admiral Lumley Lyster, and escorted by four cruisers of the 3rd Squadron and four destroyers of the 2nd Flotilla, was detached from the fleet to manœuvre independently for flying. The flying-off position was forty miles south-west of the island of Cephalonia, some 170 miles south-east of Taranto.

The crews had all been briefed and knew what they were likely to face. Some of them were not at all optimistic about their chances. Lt M.R. Maund, later writing (in *A Taranto Diary*) of Gerry Bayly, another pilot in *Eagle*'s 813, noted that 'He is always more nervy than most before a raid, but this is the worst state I have found him in yet. A sort of horror is possessing him—I can hear it in his voice—not the fear of a coward, but rather the torturings of an imagination that has had enough to digest already; the helplessness one suffers when night-mare fancies crowd your mind.'

The first strike of twelve Swordfish, led by Lt-Cdr K. ('Hooch') Williamson, Commanding Officer of 815, with Lt Norman ('Blood') Scarlett as his observer, in Swordfish L4A (*Illustrious*'s Swordfish numbers had the prefix letter 'L', *Eagle*'s 'E'), took departure at 8.57 p.m. and set course for Taranto. Six Swordfish were armed with torpedoes, four with bombs, and two with flares and bombs. The crews had trained very hard for this operation. They were as fully fit, mentally and physically, as they could be. But the Swordfish were unescorted. Darkness was their only defence. Each man, sitting in his own cockpit, was alone with his own thoughts and fears of what lay ahead.

It was a fine night, with a three-quarter moon, and some layers of thin cloud at about 8000 feet. Maund, flying E4F, one of the torpedo-droppers, reached 6000 feet. 'God! How cold it is here! The sort of cold that knows nothing of humanism and fills you until all else is drowned save perhaps fear and loneliness. Suspended between heaven and earth in a sort of no-man's land—to be sure, no man was ever meant to be here—in the abyss which men of old feared to meet if they ventured to the ends of the earth. Is it surprising that my knees are knocking together?'

The reception for the Swordfish was already warming up. Taranto harbour was protected by balloons above and torpedo nets below, and surrounded by guard-ships and shore batteries mounting a formidable barrage of AA fire. Lt Ian Swayne, one of the torpedo-droppers in L4M, made better speed to the target than the rest and his arrival, some twenty minutes early, had alerted the defences. From seaward, Maund noticed 'some quaint-coloured twinkling flashes like liver-spots have appeared in the sky to starboard. It is some time before I realise their significance; we are approaching the harbour, and the flashes are HE shells bursting in a barrage on the target area.'

The torpedo bombers had been briefed to attack the battleships lying in the outer harbour, the Mar Grande, while the bombers simultaneously attacked the cruisers and destroyers in the inner dock, the Mar Piccolo. Soon after 11 p.m., the first sequence of flares drifted down the eastern side of the Mar Grande and Williamson led the way by their cold, ghostly light, diving from 8000 to 4000 feet and finally fleeing over the harbour at thirty feet. Williamson was engrossed in his task of flying but had thoughts of Scarlett, behind him, and all observers in the same predicament. 'For him,' Williamson was quoted in Newton and Hampshire's *Taranto*, 'it must have been rather like being a passenger in a car without any brakes careering down a steep hill with a learner driver at the wheel!'

Getting his bearings from a tall factory chimney, Maund approached over the town of Taranto itself, noticing its darkened squares and the neat pattern of the streets, before heading out over the Mar Grande. In spite of the intense fire from the cruisers and the Mar Piccolo batteries, Maund closed his target. 'And so we jink and swerve, an instinct of living guiding my legs and right arm; two large clear shapes on our starboard side, are monstrous in the background of flares. We turn until the right-hand battleship is between the bars of the torpedo sight, dropping down as we do so. The water is close beneath our wheels, so close that I'm wondering which is to happen first—the torpedo going or our hitting the sea—then we level out, and almost without thought the button is pressed and a jerk tells me the "fish" is gone.

'We are back close to the shore we started from, darting in and out of a rank of merchant ships for protection's sake. But our troubles are by no means over; for in our dartings hither and thither we run slap into an *Artigliere* class destroyer. We are on top of her fo'c'sle before I realise that she hasn't opened fire on us, and though I am ready for his starboard pom-pom he has a sitting shot at something between 50 and 100 yards. The white balls come scorching across our quarter as we turn and twist over the harbour; the cruisers have turned their fire on us again, making so close a pattern that I can smell the acrid smoke of their tracer. This is the end—we cannot get away with this maelstrom around us. Yet as a trapped animal will fight like a fury for its life, so we redouble our efforts at evasion. I am thinking "either I can kill myself or they can kill me", and flying the machine close down on the water, wing-tips all but scraping it at every turn, throttle full open and wide back. With a shock I realise that we are clear of the worst of it anyway.'

One by one the torpedo-bombers jinked and swerved at masthead height across the harbour, the flares' lights outshone by the hail of tracer fire, with 'red, white and green onions streaming past the cockpits'. Lamb, in L5B, one of the flare-droppers, had a ring-side seat and recaptured the moment in one of the best thumbnail eye-witness sketches of Taranto in *War in a Stringbag*. 'Cruising along quietly at about 5000 feet, waiting for Kigell [in L4P] to begin the flare-dropping, I realised that I was watching something which had never happened before in the history of mankind and was unlikely to be repeated ever again. It was a "one-off" job. 815 squadron had been flying operationally for nearly 12 of the 15 months of the war, and for the last six months, almost without a break, we had attracted the enemy's fire for an average of at least an hour a week; but I had never imagined anything like this to be possible. Before the first Swordfish had dived to the attack, the full-throated roar from the guns of six battleships and the blast from the cruisers and destroyers made the harbour defences seem like a side-show; they were the "lunatic fringe", no more than the outer petals of the flower of flame which was hurled across the water in wave after wave by a hot-blooded race of defenders in an intense fury of agitation, raging at a target which they could only glimpse for fleeting seconds.'

After only twenty-three minutes, the attack was over. The last Swordfish pulled up and away, leaving behind Williamson and Scarlett, who had been shot down. They were picked up by an Italian destroyer and made prisoners of war. As Maund climbed away he yelled to his observer 'My Christ, Bull! Just look at that bloody awful mess—look at it! Just look at it!' A huge weeping willow of coloured fire showers over the harbour area; above it still the bursting HE shells and sprays of tadpole-like fire, whilst every now and then a brilliant flame bursts in the sky and drifts lazily down.

'At last we are free to climb. At 3000 feet it is cool and peaceful, a few shining clouds casting their dark shadows on the sea, the warm orange cockpit lights showing up the instruments that tell me all is well.'

The second strike of nine Swordfish, led by Lt-Cdr J.W. Hale, Commanding Officer of 819, began to take off at 9.23 p.m. and took departure from the ship at 9.34. Five Swordfish had torpedoes, two had bombs, and two flares and bombs. L5F, with Lt Clifford and Lt Going, had an accident on deck and took off later than the rest. L5Q, flown by Lt Morford, inadvertently jettisoned its long-range tank shortly after take-off and Morford had to return and land on, which he did safely, in spite of some brisk AA fire from *Illustrious*'s escorts.

Meanwhile the surviving seven, with Clifford trying to catch them up, could see Taranto from sixty miles away. Lt Sutton, Torrens-Spence's observer in L5K, one of the torpedo-droppers, wrote in *Taranto*: 'I gazed down upon a twinkling mass of orange-red lights which I knew was a solid curtain of bursting shells through which we had to fly. It looked absolutely terrifying.'

Their attack fully lived up to Sutton's expectations. 'Down, down we went in that screaming, whistling torpedo dive. All the enemy close-range weapons had now opened fire. We could see multiple batteries by the entrance to the inner harbour pouring stuff out right next to our dropping position. Tracer and incendiaries and horrible things we called "flaming onions" came streaming up at us.' In the middle of his attack, Torrens-Spence had to take drastic avoiding action to avoid E4H, Bayly's Swordfish. Bayly's premonitions were being borne out. Nobody saw him or his observer Tod Slaughter alive again.

Torrens-Spence's target 'saw us and opened fire. The flash of her close-range weapons stabbed at us first one and then another along her length opened up. We were coming in on her beam, and we were the centre of an incredible mass of cross-fire from the cruisers and battleships and shore batteries. No worries about clear range or gun zones for the Italians. They just fired everything they had except the 15-inch, and I could see the shots from the battleships bursting among the cruisers and merchant ships. The place stank of cordite and incendiaries and was wreathed in smoke.'

One by one the Swordfish survivors returned to the deck. Maund's experience was typical. 'A crowd of people in the glaring AIO [Action Information Office] jabbering like men possessed—as indeed they are. I am giving my evidence. A Littorio battleship, I think. Yes, good clear run of 1000 yards. A good drop. Oh—about thirty feet. The wardroom warm and sane. Three rapid whiskies-and-soda, followed by some eggs and bacon I can scarcely taste. A camp-bed on the quarterdeck in the grey light of dawn. Unconsciousness.'

Photo-reconnaissance showed that the results of the raid went beyond anybody's wildest dreams. The battleship *Littorio* sustained three torpedo hits which put her out of action for nearly a year. The older battleship *Caio Diulio* had one hit, but so serious was the resultant flooding that she had to be hurriedly beached to stop her foundering. A third battleship *Conte de Cavour* also had one torpedo hit which damaged her so badly that she sank in shallow water. She was raised eventually, but took no further part in the war. The heavy cruiser *Trento* had one bomb hit and there was also damage to some destroyers, a seaplane base and oil storage tanks beside the Mar Piccolo. All this for the loss of two Swordfish, of which the crew of one were rescued. When *Illustrious* rejoined the fleet, Cunningham made her one marvellously laconic signal: 'Manœuvre well executed.'

The battle at Taranto was, as Lord St Vincent said of his own battle in an earlier century, 'a victory very essential to England at this moment'. It changed the strategic balance of capital ships in the Mediterranean, literally overnight. As Captain Boyd told *Illustrious*'s ship's company 'This will cheer the entire free world.' Cunningham himself said that 'as an example of "economy of force" it is probably unsurpassed.' But it was typical of Cunningham that he wished to pursue and totally destroy an already beaten foe. He was disappointed that no second strike could be carried out the following night. With hindsight, perhaps it was just as well. The Swordfish could not have expected to get away so lightly again. But the weather was judged unsuitable or maybe Captain Boyd believed his aircrews had done enough, and the second strike was cancelled. One had been enough. It was no wonder that Captain Bragadin, who was on duty that night in the operations room of the Italian Ministry of Marine, later wrote 'It was as if we had lost a great naval battle, and could not foresee being able to recover from the consequences.'

It was a great pity that such a sweet victory should have been somewhat soured by the manner in which the awards were given and announced. *Illustrious*'s ship's company confidently expected that Williamson and Hale would both be given the Victoria Cross, and every man who took part in the strike would be handsomely decorated. On 20 December, Williamson and Hale were both awarded the DSO, while their respective observers, Scarlett and Carline, got the DSC, as did Patch and his observer Goodwin. Nothing else. The irate sailors of *Illustrious* are supposed to have torn down the notices giving the details of the decorations from the ship's notice-boards.

In the New Year's Honours List, Admiral Lyster got the CB, Boyd and Bridge both CBEs. In May 1941, six months after the events at Taranto, DSOs were awarded to Clifford and Going, DSCs to thirteen others, and eighteen more, including Maund, Bayly and Slaughter, were mentioned in despatches. Not one officer or rating of the maintenance and non-flying personnel in *Illustrious* received any recognition at all. 'This arm,' wrote Captain Boyd in his official despatch, 'has had a long struggle with adverse opinions.'

Dividends from the Taranto incident were not long in coming. Capital ships could be spared from the Mediterranean fleet for other theatres, while Cunningham's ships could range the Mediterranean ever more freely. On 25 November, for instance, *Illustrious*'s aircraft raided targets in the Dodecanese almost at the same time as *Eagle*'s Swordfish were over Tripoli, more than 750 miles to the west, while on 27 November off Cape Spartivento, in southern Sardinia, a powerful Italian force which included two battleships withdrew from a tactical position favourable to them, on the mere threat of attack from *Ark Royal*'s aircraft.

3

The Long Year of 1941

The Luftwaffe's Revenge

The Taranto incident had another effect, most unfortunate for *Illustrious* herself: it brought the Luftwaffe to the Mediterranean for the first time. Dismayed by defeat, the Italians asked the Germans for assistance and the Germans, already exasperated by their flabby ally's performance, responded. By the end of 1940 between 300 and 400 Junkers Ju87 and Ju88 bombers of General Geissler's *Fliegerkorps* X had arrived at airfields in southern Italy and Sicily. The *Kriegsmarine* had absorbed the lessons of Norway, in which *Konigsberg* had been sunk, and the main Allied air effort had been mounted from carriers. Geissler's aircrews had therefore been specially trained for anti-shipping attacks.

On 10 January 1941 *Illustrious* was at sea, off the island of Pantelleria, west of Malta, providing air cover for the fleet and for a convoy to Greece. She was flying anti-submarine Swordfish patrols and keeping a CAP of Fulmars overhead. Admiral Lyster and Captain Boyd had both represented to Cunningham that there was no need for their ship to steam so close to enemy airfields. She could still successfully provide cover for the fleet from longer range (although she had only six Fulmars serviceable). Cunningham replied that the mere sight of *Illustrious* was good for morale.

The Fulmars shot down one shadower, but there were others and it seems that *Illustrious*'s position was accurately reported, for at 1.30 p.m. two Savoia S79s made a torpedo attack. Their torpedoes were avoided but unfortunately the Savoias drew the Fulmar CAP down low just at the moment when Stuka dive-bombers clustered overhead. From the moment the Stukas tipped over, one by one, from their 'push-over' height of about 11,000 feet, it was clear they were concentrating upon the carrier.

It was also clear that a new foe was now in the field. This was flying of a skill and aggressiveness which Cunningham's ships had never experienced before. Cunningham himself who was on *Warspite*'s flag bridge, later wrote in his memoirs *A Sailor's Odyssey* that he was too interested in this new form of dive-bombing attack really to be frightened. He said he had no doubt they were watching complete experts. Formed roughly in a large circle over the fleet the Stukas peeled off one by one when they reached an attacking position. Cunningham said he could not but admire the skill and precision of it all.

The slow-climbing Fulmars of the CAP clawed their way upwards, and three fresh Fulmars were launched, but they were not in time to intercept and *Illustrious* had to rely upon her own and the fleet's AA gunfire. It was not enough. For minutes at a time *Illustrious*'s hull and island superstructure were completely hidden in the leaping curtains of water thrown up by near-misses. Eventually one direct hit from a 500–kg bomb penetrated the armoured flight deck, blew out the after lift, either destroyed or damaged every aircraft in the hangar, killed or wounded most of the men there and started numerous fires. Other direct hits, six in all, with three near-misses, utterly smashed the after lift, crippled the ship's steering gear and wrecked the after ammunition hoists. Blast reached the level of the wardroom and cabin decks and reduced the after part of the ship to black silence, lit by fires raging all along the main deck. But for her armoured flight deck, *Illustrious* must have been lost. Fortunately, she was still able to steam, and steering by main engines (altering the revolutions on the shafts) Captain Boyd headed the ship for Malta. Below, the officers and ship's company fought fires, gathered in the dead, dying and wounded, and began to pump out flooded compartments.

Illustrious's Fulmars and surviving Swordfish flew to Malta where, later in the day, their ship followed them. After several more bombing attacks in the following week, while she lay in Valetta harbour, *Illustrious* managed to slip away on the evening of 23 January, and escaped to Alexandria. Eventually she steamed through the Suez Canal and reached Norfolk, Virginia, for a refit which virtually meant rebuilding half the ship.

Malta

In Malta, island of bells and smells, and of bombs in 1941, *Illustrious*'s surviving aircraft went to Hal Far where 830 Squadron, one of the most remarkable units in the Fleet Air Arm, had been operating for some months. It had originally been 767 Squadron, formed from the survivors of *Courageous*'s 811 and 822, doing deck landing training on *Argus*, based at Hyéres in the south of France. On 14 June, after Italy entered the war, the squadron improvised its own bombing raid. Nine Swordfish, led by Lt-Cdr G.C. Dickins, took off with French bombs (fastened to the racks with spunyarn and fused by hand) and dropped twenty-three of them somewhere over Genoa.

On 18 June, the whole squadron of 24 Swordfish flew across the Mediterranean to Bone, in North Africa, taking some of the ground crews with it, and the rest going in a commandeered ship. From Bone, half the squadron flew home via Casablanca and Gibraltar, but the other twelve, the striking half, flew via Medjej-el-Bab to Malta, where they began to operate under the RAF against targets in Sicily.

The squadron kept two Swordfish always standing by at fifteen minutes' notice for anti-submarine searches, and, as the weeks passed, began to diversify—one day bombing Palermo, or oil tanks at Augusta, another day laying mines, a third carrying out torpedo attacks on an enemy convoy. Torpedo attacks became the squadron speciality. After RAF reconnaissance aircraft had sighted the ships between Sicily and Cape Bon, the Swordfish would take off from Malta, timing their attack to arrive just after dusk. Their main danger by day was the standing patrol of Messerschmitt Me109s, which circled constantly over Malta, waiting for the returning Swordfish. The Swordfish were in any case antique training types, of 1935 vintage, with no blind-flying instruments or long-range tanks. But by July 1941, instrument panels had been fitted and Air Surface Vessel (ASV) radar.

'Life on the island was not pleasant,' wrote Leading Airman (TAG) Nat Gold in *The Knights of Malta*. He joined the squadron on 10 January 1941. 'At times we were short of food, torpedoes and shells, the latter having to be rationed. There was continuous bombing with screaming bombs, and raids of over 100 aircraft at a time on one target (during April 1942 750 tons of bombs fell on Hal Far alone)—and then all hands spread across the drome to pick up shrapnel or fill bomb craters. Our living quarters were so badly damaged, we were evacuated to a house in the village of Zurrieq, which we shared with rats, mice and fleas.

'We normally flew during the period when the moon was up, taking off at midnight, returning around 6–7 a.m. the following morning, after which we were debriefed, invited to have breakfast in the Officers Mess, then turn-to to service our aircraft. This was a difficult job, the aircraft being dispersed, and Jerry keeping up continuous alerts. At noon we were given a make-and-mend, and it was advisable to sleep in the air-raid shelter during the afternoon. At the end of approximately a fortnight, the Squadron stood down and the air gunners went to a rest camp for a few days at Ghain-Tuffieha, and ate, swam, and got gloriously drunk.' The Squadron also had several RAF personnel including one Leading Aircraftman Hendry who, Nat Gold said, 'was mad—laid on the tarmac at Luqa and took photographs whilst Jerry bombed the place'.

830 had several brilliant successes, including one convoy strike on 23 January 1942 when six Swordfish, led by Lt-Cdr F.H.E. Hopkins, battled through a raging north-easterly gale (the *gregale*) and driving rain to get two hits on the 24,000-ton Lloyd-Triestnoy liner *Victoria*. Hopkins himself flew for a total of nearly twelve hours that day in conditions where normally flying would have been thought impossible. He was awarded an immediate DSO.

But it was not all success. 'One pilot, a Sub-Lt RNVR,' wrote Nat Gold, 'tried but failed. He "chickened out" on each operation he attempted. On his last flight, on 11 November 1941, he flew with me as I had gained some experience by now. (Experience allowed you to fly with the Commanding Officer on one operation, and the most "goon" pilot on the next, to give him confidence). I was not at all happy; one of the escorts of the convoy was a cruiser, and, knowing the prospects of this pilot, I felt our chances would be very slim. However, after one and a half hours he turned back. We were the only aircraft to return. The rest of the Squadron, six Swordfish, together with the shadowing aircraft, (an RAF Wellington), never returned. What remained of the squadron received sympathy from Churchill and all the other big-wigs. This pilot left the island on the next available aircraft. I felt extremely sorry for him, yet grateful for probably saving my life. The whole operation should never have happened.'

In October 1941, 830 were joined by 828, flown off from *Ark Royal*. In spite of their casualties and their aircraft losses—at times they were down to a couple of aircraft—the damage these squadrons and the RAF did to the enemy has in the end to be rendered into figures: from May to November 1941, 110,000 tons of shipping sunk and 130,000 damaged. By the end of 1942, over 400,000 tons sunk, including fifty-three ships of nearly 40,000 tons in the last five months of 1942 alone.

At times the enemy must have thought that Swordfish were everywhere. They operated against Axis shipping from a secret base at Paramuthia, in the mountains of Albania. In May 1941 they played their part in suppressing Raschid Ali's revolt in Syria, and at 3 a.m. on 16 June, north of Cyprus, five Swordfish torpedoed and sank the Vichy destroyer *Chevalier Paul*, by night, when she was steaming at 35 knots, carrying reinforcements from Salonika to Syria.

In March 1941 *Eagle*'s Swordfish squadron flew to Port Sudan to support the army in the final advance to the Red Sea against the defeated Italians. One day about ten miles off Port Sudan, Cdr Keighly-Peach suddenly saw one of his Swordfish diving towards the sea. 'On following down through the clouds,' he said in *Fleet Air Arm*, 'I saw a lovely fat Italian destroyer right beneath me, and a second later another zigging away ahead of her. They fired at us extremely ineffectively and after dropping my stick, which fell across the stern of the leading ship without hitting, I made all speed back to Port Sudan to whip up the main party.'

When the 'main party' was whipped up, they bombed the destroyers in relays, as fast as they could refuel and return; they sank two destroyers, while another two were beached and abandoned, to be later destroyed by the destroyer *Kingston.* A fifth was scuttled in Massawa harbour. *Kingston* later presented *Eagle* with the Italian ensign from one of the beached destroyers and it was ceremonially flown from a gaff inside the hangar.

Matapan

While *Eagle* went east through the Canal to be employed on trade protection in the Atlantic, having played a valuable and honourable part in the Mediterranean, the new aircraft carrier *Formidable* (Captain A.W. La T. Bisset) came through the Suez Canal to join the Mediterranean Fleet on 10 March 1941, having been delayed for a time by mines laid in the canal. The second of the *Illustrious* Class, with an armoured flight deck, she had twelve Fulmars of 806 Squadron, and twenty-one Albacores of 826 and 829 Squadrons embarked. The Fairey Albacore was another three-seater torpedo-bomber-reconnaissance biplane, similar to the Swordfish which it was designed to supercede (but in fact it was withdrawn from service in 1943, before the Swordfish). It had a bigger engine, some 30 more knots top speed, increased range, improved instrumentation and an enclosed cockpit. As they said, 'An effort has been made to suggest you are travelling first-class in an Albacore.'

On 27 March, a patrolling Sunderland of 201 Group RAF reported units of the Italian fleet at sea, south-west of Crete. It seemed that the Italian Supermarina had at last yielded to German pressure to take some action against the flow of Allied convoys from Egypt to support the campaign in Greece. The Mediterranean battle fleet, of *Warspite, Valiant* and *Barham*, sailed from Alexandria that evening, with *Formidable* in company, flying the flag of the newly promoted Rear Admiral Denis Boyd, Rear Admiral, Aircraft Carriers. She had less than her full complement of aircraft, with thirteen Fulmars, of 806 and some survivors of *Illustrious*'s 803, ten Albacores and four Swordfish of 826 and 829.

In the action off Cape Matapan which followed the next day, 28 March, naval aircraft actually did what they had threatened to do off Cape Spartivento and succeeded in slowing down the enemy's ships, until the battle fleet could catch up. In fact, had they scored more torpedo hits, the aircraft would have demonstrated that they were the ultimate striking weapon on their own.

The fleet's movements that day were accompanied by the most intense air activity, both carrier and shore-based, that the Royal Navy had so far ever experienced in any of its actions. But while Cunningham received a stream of air intelligence (admittedly not all of it accurate) and was able to command air strikes when he wished, his opponent, the Italian Admiral Iachino, had no strike aircraft under his own hand, and he later complained bitterly and quite justifiably that he was given almost no air reconnaissance information and no fighter cover of any kind at any time, while his own ships were both sighted and attacked by various types of Allied aircraft at various times.

Formidable flew off her first search at 5.50, soon after dawn on the 28th, when the fleet was about 150 miles south of the eastern end of Crete. At 7.20 a.m. the searchers reported two formations of cruisers and destroyers, about 100 miles to the north-west, off the southern coast of Crete. In the flagship *Warspite*, hopes began to rise. At the prospect of action Cunningham began his customary 'caged tiger act', walking furiously up and down his flag bridge.

For a time it was feared that one of the forces reported might be Vice Admiral H.D. Pridham-Wippell's cruisers and destroyers but at 7.45 a.m. all doubts vanished when look-outs in the cruiser *Orion* sighted smoke to the north-west. It was the enemy, and in the classical mode of light forces, Pridham-Wippell's ships retired, hoping to entice the enemy after them on to the guns of the battle fleet.

This time, the bait was taken up rather too enthusiastically, for the Italian force included a battleship and Pridham-Wippell's ships soon found themselves fleeing south-east in deadly earnest, with *Vittorio Veneto* firing at them from the port quarter and three 10,000 heavy cruisers on the starboard quarter. *Warspite* had intercepted, and caught the urgent note, of Pridham-Wippell's signals to his ships: 'Make smoke by all available means', 'Turn together to 180 degrees' and 'Proceed at your utmost speed'. Cunningham interrupted his staff's speculations about their meaning, saying 'Don't be so damned silly! He's sighted the enemy battle fleet. . . .'

They had indeed, and matters might have gone hard for Pridham-Wippell and his ships. *Vittorio Veneto* had their range and her salvoes were dropping ever nearer, while the very fast Italian cruisers were steadily overtaking. Luckily, the Italians were distracted by *Formidable*'s first torpedo strike of the day, of six Albacores of 826 led by Lt-Cdr Gerald Saunt, escorted by two Fulmars. They had been launched at 10 a.m. and arrived at about 11.25, in the nick of time. Lt F.H.E. Hopkins, Saunt's observer, could see Pridham-Wippell's predicament at once. 'It now became obvious' he said in Captain S.W.C. Pack's book on Matapan 'that unless we could do something quickly our cruisers would be picked off one by one at long range by the *Vittorio*. The trouble was that we were all abaft of *Vittorio*, she was steaming at 30 knots, the wind at our height was 30 knots against us, so that since our air speed was only 90 knots we were catching up at a relative speed of only 30 knots. I think it took the best part of twenty minutes to creep up to a suitable attacking position ahead of *Vittorio*.'

At 11.27 the first three Albacores dropped their torpedoes off *Vittorio*'s starboard bow and when she turned to comb the tracks, the second three attacked from the port bow. The pilots were convinced they had hit, but in fact all six torpedoes missed, two ahead, four astern. But they made Iachino break off the action and retire to the north-west, pursued by the two Fulmars who diverted the battleship's gun crews attention by strafing her upper deck. Lt Donald Gibson of 803 also shot down a harassing Ju88. To add further to Iachino's troubles, three of *Illustrious*'s Swordfish of 815, flying from Maleme in Crete, made a torpedo attack on the heavy cruiser *Bolzano* at 12.05. All their torpedoes missed, and the aircraft evaded AA fire and escaped.

As Iachino's ships turned for home, Cunningham's ships were only forty-five miles away to the south-east and steaming at top speed in pursuit. Top speed, however, was only 22 knots because *Warspite* had got sand in one of her main condensers from a shallow bank while leaving Alexandria. Even after the fleet engineer officer had gone below to have a few words with the ship's staff, her speed only crept up to 24 knots. The enemy had to be slowed down and so, at 12.20 p.m. another strike of three Albacores and two Swordfish of 829, led by Lt Cdr Dalyell-Stead, with two escorting Fulmars, was launched. Saunt's returning strike was then landed on. The last aircraft was barely on when two Savoia-Marchetti SM79s made a good attack on *Formidable*, whose violent manœuvring managed to evade the torpedoes.

Dalyell-Stead's strike sighted the enemy at 3.10 and nine minutes later carried out their attack which Iachino himself said was 'conducted this time with particular ability and bravery in aircraft which had evidently come from an aircraft carrier'. The three Albacores approached *Vittorio Veneto* in the middle of a bombing attack by RAF Blenheims of 113 Squadron flying from Greece (making their second attack of the day), and were able to get close almost unmolested before being sighted.

Dalyell-Stead closed to within 1000 yards before dropping, while every gun on *Vittorio Veneto*'s port side concentrated on him. After launching his torpedo Dalyell-Stead broke off to his port hand, as

THE ENGLISH CHANNEL AND THE MEDITERRANEAN
North Sea
DENMARK
SWEDEN
EIRE
GREAT BRITAIN
Dover
Ostend
HOLLAND
GERMANY
ENGLISH CHANNEL
Boulogne
Dunkirk
BELGIUM
Cherbourg
Brest
LUX.
Bay of Biscay
FRANCE
SWITZ.
AUSTRIA
Bordeaux
ITALY
Rimini
Pisa
Leghorn
Marseilles
PORTUGAL
SPAIN
CORSICA
La Maddalena
BALEARIC ISLANDS
SARDINIA
Naples
Decimomannu
Tyrrhenian Sea
Cagliari
SICILY
Palermo
Gibraltar
MEDITERRANEAN SEA
C. Bon
TANGIER
Blida
Bizerta
Trapani
Catania
Skerki Bank
The Narrows
Syracuse
SP. MOROCCO
Oran
Algiers
Bône
Tunis
Comisa
Casablanca
PANTELLERIA
MALTA
LAMPEDUSA
TUNISIA
Sfax
KERKENAH IS.
MOROCCO
ALGERIA
Tripoli
0
500
Miles
TPIPOLITANIA

~ARTHUR BANKS~
LITHUANIA
Baltic Sea
EAST PRUSSIA
U.S.S.R.
POLAND
CZECHOSLOVAKIA
HUNGARY
RUMANIA
Black Sea
YUGOSLAVIA
BULGARIA
ADRIATIC SEA
ALBANIA
Bosporus
GREECE
Valona
Dardanelles
TURKEY
Taranto
Aegean Sea
Ionian Sea
Athens
Piraeus
Aleppo
Reggio Calabria
Messina
Navarino
SYRIA
RHODES
Maleme
C.Matapan
CYPRUS
DODECANESE Is.
CRETE
Haifa
MEDITERRANEAN
SEA
PALESTINE
TRANS-JORDAN
Derna
Gazala
Tobruk
Sidi Barrani
Mersa Matruh
Alexandria
Port Said
Bomba
Benghazi
Bardia
Sollum
Suez Canal
Suez
Buerat
El Alamein
Cairo
Gulf of Sirté
CYRENAICA
EGYPT
El Agheila
LIBYA
ARABIA

though hoping to reach *Vittorio Veneto*'s starboard side where the AA fire seemed less intense. In so doing he presented an almost perfect plan silhouette of his Albacore to the Italian gunners and his aircraft was shot down into the sea. Dalyell-Stead, Lt Cooker, his observer, and PO Blenkhorn, his TAG, were all lost.

As the great ship turned to starboard, the two Swordfish attacked from that side, while the Fulmars carried on distracting *Vittorio Veneto*'s gun crews with strafing runs along her upper deck and superstructure. Meanwhile, Lt A.S. Whitworth, the Squadron second-in-command, who took over from Dalyell-Stead, watched his Commanding Officer's torpedo run. 'The column of water off the port quarter was seen by nearly all crews,' he reported, 'I consider it a hit.'

Whitworth was quite right. The torpedo hit *Vittorio Veneto* just above her port outer screw. Hundreds of tons of water flooded into the breached compartments. A great tower of water crashed down on to the quarterdeck. A Blenheim scored an almost simultaneous near-miss at that moment and huge, dark smoke rings belched, as if in outrage, from the battleship's funnel. Her speed fell away, and by 3.30 she had come to a dead stop in the water, listing to port and settling by the stern. Lt Haworth, high above in a shadowing Swordfish, signalled to *Warspite* at 3.58 p.m. 'Enemy has made a large decrease in speed.'

In *Warspite*, Cunningham redoubled his gleeful pacing up and down. *Now*, at last, his enemy was delivered into his hand, so he thought. But it was not to be. *Vittorio Veneto* got under way again, worked up to 16 knots and later to 19, chivvied as she went by further Blenheim attacks by 84 Squadron RAF. They scored only a near-miss on the cruiser *Garibaldi* but these constant attacks by various types of aircraft, seemingly every hour on the hour, served to increase Iachino's sense of grievance and, possibly, affected his judgement of his own aircraft.

Formidable was ordered to launch another strike at maximum strength at dusk. Six Albacores and two Swordfish began to fly off at 5.35 with orders to land on Maleme after their attack. Meanwhile, from Maleme, two Swordfish led by Lt Torrens-Spence also took off to attack the Italian fleet.

The Italian look-outs sighted the aircraft approaching from astern about fifteen minutes before sunset. 'These were the planes,' said Iachino, 'whose job it was to give us the *coup de grace* at nightfall.' High above, in the shadowing float-fitted Swordfish from *Warspite*, was the fleet observer, Lt A.D. Bolt, who throughout the day gave a superb exhibition of observing and reporting, demonstrating to all, new aircrew and ship's staff officers alike, just how a fleet observer should conduct himself. 'At sunset' Bolt observed, 'the Italian Fleet formed itself into a compact mass which made it a difficult target for the dusk torpedo attack. I saw the attack develop from a position about five miles astern of the Italian fleet. . . . The attack was most spectacular, the Italian fleet pouring out vast quantities of coloured tracer from their close-range weapons.'

By 7.30 it was practically dark. The rearmost destroyer caught a glimpse of an attacker and opened fire. The next in line followed suit. Every ship was already at action stations with gun crews keyed up for the attack, and the uproar spread from ship to ship, until the whole of the Italian force closely clustered around *Vittorio Veneto* was contributing to a magnificent display of tracer and bursting explosive, with a background of sweeping searchlights and great dark rolling clouds. Hopkins was once more Saunt's observer: 'When we eventually went into the attack from the dark side with the Italians silhouetted against the last glow of light in the west, we found that we had been spotted at long range and were met with an impassable barrage of fire. We were forced to withdraw, and split up and came in again individually from different angles. The barrage of fire put up by the Italians was immensely spectacular but not very effective. A good deal of hose-piping went on which resulted in a number of their ships hitting each other but little damage to our aircraft.'

The last of *Formidable*'s aircraft, Albacore 5A, piloted by Sub-Lt C.P.C. Williams, dived on the 8000-ton cruiser *Pola*, the middle cruiser on *Vittorio Veneto*'s starboard wing. His torpedo hit at 7.46 and brought *Pola* to a stop. Iachino was unaware of the loss and steamed on. The aircraft flew back with, it seemed, little to report. Torrens-Spence (whose torpedo hit on *Pola* it might have been) reported no results observed. For all that sound and fury, one hit on a cruiser did not appear very much.

In fact, that hit had a very important result. When Iachino heard that *Pola* had dropped out, he sent the heavy cruisers *Zara* and *Fiume* and two destroyers back to support her. Iachino decided, on the basis of radio direction-finding bearings, that Cunningham's heavy ships could not be less than 170 miles away. He ignored aircraft sighting reports which placed his enemy much nearer. In an action later that night, Cunningham's battleships came upon the Italian cruisers, and sank *Zara* and *Fiume*. *Pola* was dispatched by torpedo the next morning. The two destroyers, *Alfieri* and *Carducci*, were also sunk. The Fleet's aircraft had done for Cunningham exactly what he had hoped, and expected, they would do.

The Matapan affair was a tremendous victory but the Luftwaffe took their revenge on *Formidable*, as they had done on *Illustrious*. On 26 May, during the battle for Crete, *Formidable* launched a dawn strike of four Albacores and four Fulmars to bomb and strafe the airfield at Scarpanto, an island east of Crete, where many of the German aircraft harrassing Allied troops in Crete were based. At 1.25 that afternoon *Formidable* was found and attacked by some thirty Stukas flying, not from Scarpanto, but from North Africa. By a coincidence these Stukas were from II Group of the 'Immelmann' *Stukageschwader* (Dive Bomber Wing) which had also attacked *Illustrious*, and soon they had inflicted almost the same damage on *Formidable*. After two direct 500-kg bomb hits and one very near-miss, *Formidable* followed *Illustrious*'s melancholy track, back to Alexandria, through the Suez Canal and across to the United States for major repairs.

Force H

After *Formidable*'s departure, no fleet carrier operated regularly in the eastern Mediterranean for another two years. Air cover for the fleet was provided by No. 201 Naval Co-operation Group of the Middle East Air Force, with disembarked naval squadrons as part of their aircraft strength. With the best will in the world and the utmost desire to co-operate, no aircraft operated by the RAF could ever give Cunningham the quick response of a carrier under his own command. In the evacuation of Crete, the Navy knew that a terrible price would have to be paid, deprived as they were of their air cover. And so it was, with a long list of warships sunk or damaged by air attack. But the Army had to be taken off, come what may. As Cunningham himself said, it was a matter of honour. 'It would take three years to build a new ship. It would take three hundred to rebuild a tradition.'

While Cunningham's ships endured their ordeal in the east, Force H held the ring in the western Mediterranean, once more showing the versatility of naval aircraft. In February 1941, *Ark Royal*'s aircraft attacked the Tirso Dam in northern Sicily and bombed targets in Genoa. In March, one of *Ark Royal*'s Swordfish sighted *Scharnhorst* and *Gneisenau* in the Atlantic, but a faulty radio set and very poor visibility prevented an attack. In April and May *Ark* joined *Argus* in ferrying badly needed fighters to Malta, carrying them to within range of the island and then flying the fighters off, with a naval Skua to guide them.

Argus seemed too lady-like for war. It seemed almost ungentlemanly to bomb her. As one of her engineer officers wrote in *The Naval Review* in February 1945: 'The sight of the *Argus*, firing a pathetic hail of lead from her ancient pieces of ordnance into an assortment of hostile aircraft, weaving wildly to avoid their bombs or torpedoes, flat out and emitting sparks and showers of soot from her funnel ducts and then paddling sedately on her way when it was all over, was enough to provoke paroxysms of mirth in the breasts of the onlookers. On these occasions she was like nothing so much as an old lady shocked into indelicate exposure while crossing a busy street and letting down her skirts with a sigh of relief on safely gaining the pavement. Shaken; but, for the time being, secure.' In fact, due to the skill of successive captains and the efficiency of 'Weary Willie', her patent circa 1908 'steam tiller', which enabled the ship to 'turn on a sixpence' *Argus* came through the war unscathed.

In May 1941, within six days of carrying out a ferrying run to Malta, *Ark Royal* played her part in the drama of the *Bismarck*, more than 2000 miles to the west.

Bismarck

The destruction of the *Bismarck* would never have been possible without the actions of several arms: heavy ships, shadowing cruisers, torpedo attacks from destroyers, aircraft of Coastal Command and of the Fleet Air Arm, both shore and carrier-based. Each component played its part in the story, as though acting in a well-made play.

Bismarck and the heavy cruiser *Prinz Eugen* had sailed from Gdynia on 18 May for an extended sortie against Allied shipping in the Atlantic. When they were reported by intelligence sources on the 20th, Coastal Command mounted searches of the Norwegian coast which, on the 22nd found the two ships at Korfjord, a few miles south of Bergen. For Admiral Sir John Tovey, Commander-in-Chief Home Fleet, at Scapa Flow, it was essential to have up-to-date information of *Bismarck*'s movements, but bad weather prevented any further reconnaissance.

The problem was brilliantly solved by Captain H.L. St J. Fancourt, commanding HMS *Sparrowhawk*, the naval air station at Hatston in the Orkneys. He had already moved 829 Squadron of Albacores, based at Hatston for convoy duties with Coastal Command, up to the Shetlands where they would be just in range for a torpedo strike at the German ships. Knowing the Commander-in-Chief's anxiety, and needing information himself for the Albacore strike, Fancourt on his own initiative organised a reconnaissance by a Glenn Martin Maryland of 771 Squadron, normally used for target-towing. It was flown by Lt Noel Goddard with Cdr Geoffrey Rotherham as observer. 'Hank' Rotherham was actually appointed to Hatston for administrative duties, not for flying at all, but he was in fact one of the most experienced observers in the Navy, with many years' flying behind him.

The Maryland took off at 4.30 p.m. and at once ran into weather conditions which, as Cdr Rotherham himself said (in *Send her Victorious*) 'were really terrible, with "clag" right down to the surface of the sea. We had to keep coming down to check the wind speed for navigation and on one occasion, I saw the surface of the sea only thirty feet below the aircraft and the pilot hadn't seen it.' When they estimated they were near the Norwegian coast, they went down again to sea level, and just as Goddard and Rotherham were bracing themselves for the possibility of flying straight into a cliff, the weather cleared and there, dead ahead, was Marstein, the very island they were looking for.

Soon they found Korfjord, and searched it from end to end. They also looked into the nearby Grimstad and Kalvanes Fjords. But there was no sign of *Bismarck*. To make sure, they flew over Bergen itself and circled the harbour, amidst a fierce storm of flak. Again, there was no sign of either German ship. Fearful of being shot down before he could transmit his message, Rotherham would have been content to retreat into cloud, but was unable to get in touch through the plane's intercom with Goddard, who put the Maryland's nose hard down and thundered over Bergen only 200 feet above the roof-tops and so out to sea, all hands mightily relieved at being still alive. The Maryland's radio was defective, so PO Milne the TAG passed the message over the target-towing frequency, actually interrupting every day target-towing traffic with this crucial operational signal. The Maryland landed at Sumbrugh in the Shetlands at 7.45 p.m. when Rotherham was at once called to the telephone to talk to the Commander-in-Chief's Chief of Staff, and enlarge on his signal. At 10.15, the Home Fleet in Scapa was aweigh and one of the greatest hunts in the Navy's history was on.

Naval aircraft again took a part late on 24 May, by which time a great deal had happened. *Bismarck* and *Prinz Eugen* were sighted on the evening of the 23rd and shadowed by the cruisers *Norfolk* and *Suffolk*. The cruisers had delivered their quarry within gun range, and what they imagined would be certain destruction, of *Hood* and *Prince of Wales*. Instead, *Bismarck* shocked the Navy and the whole world by sinking *Hood* in a short engagement in the early hours of the 24th. Three survivors were picked up of a total company of 1409. *Bismarck* also hit *Prince of Wales* and forced her to break off the action. *Prince of Wales* was a new ship and hampered by machinery troubles (one of her main 15-inch guns could only join in the first salvo) but still managed to score two hits on *Bismarck*. The damage was superficial but did deprive *Bismarck* of the use of some 1000 tons of fuel. The German Admiral Lutjens decided to abandon the Atlantic venture and make for France instead. Still shadowed by *Prince of Wales*, the cruisers and by Coastal Command, *Bismarck* began a run to the south which brought her eventually within range of aircraft from *Victorious*.

Victorious (Captain H.C. Bovell), another of the *Illustrious* Class, was also new and had recently been employed on ferrying duties. She had on board only nine Swordfish of 825 Squadron and six Fulmars of 800Z Flight. Some of the aircrew had been on board only a week and others had completed only a few deck landings. Nevertheless, 825 had a most experienced and daring pilot and a determined leader in the Commanding Officer, Lt-Cdr Eugene Esmonde. Pre-war, he had years of flying experience with Imperial Airways before joining the Fleet Air Arm in May 1939, and had actually been a survivor of *Courageous*.

There was no sign that the action with *Prince of Wales* had slowed *Bismarck* down and so, on the afternoon of the 24th, Admiral Tovey detached *Victorious*, escorted by the 2nd Cruiser Squadron with Rear Admiral Alban Curteis flying his flag in *Galatea*. 825's Swordfish were now the Commander-in-Chief's only hope of catching *Bismarck*.

As the evening drew on and *Victorious* prepared to launch aircraft, *Bismarck* suddenly altered to the west and opened a short sharp gun engagement with the shadowing cruisers. This was actually to cover the withdrawal of *Prinz Eugen* which proceeded independently to Brest. But the manœuvring also served to open the range from *Victorious* whose aircraft could not intercept until later than planned, and at a range much further than Captain Bovell would have wished for his raw aircrews. However, as *Bismarck*'s range was closing slowly and the weather looked like getting worse, Admiral Curteis could not wait any longer and ordered the strike to proceed. The nine Swordfish flew off at 10 p.m., followed by three shadowing Fulmars, who would be relieved by two more, flown off two hours later.

It was still light, because sunset in those high latitudes was not until some fifty minutes after midnight. On the flight deck, men and aircraft were soaked by heavy rainsqualls and drifts of spray. The wind was freshening from the north-west and raising a long swell. Led by Esmonde, wearing the old blue 'pussers' raincoat he always wore for flying, 825 took off into a darkening sky, beneath lowering clouds. Captain Bovell watched them go, wondering how they would ever get them on board again. Some in *Victorious* wondered whether they would ever see them again.

The nine Swordfish, flying in three 'vics' of three aircraft each, headed 225°, at 85 knots. They had 120 miles to go. The visibility was variable, sometimes dropping to three miles in low cloud, but the Swordfish were fitted with ASV radar and at 11.27 p.m. obtained a radar contact. It was *Bismarck* herself, in sight about twenty miles away. Esmonde took his squadron back up into cloud while he closed the target, but when he emerged again *Bismarck* had disappeared. Esmonde led in the direction of the enemy, wrongly as it happened, and was corrected by frantic flashing from the cruiser *Norfolk*, who gave him the right course to steer. Soon Esmonde had another radar contact. This one had to be *Bismarck* and the Swordfish dropped out of the clouds on their attacking runs. To their dismay, they found the 'target' was actually the US Coastguard cutter *Modoc*. But *Bismarck* herself was only six miles away, her gunners now fully and furiously alert and firing. All surprise had been lost.

The Germans themselves were amazed at being attacked by such antique aircraft, 'to see such obsolete-looking planes, having the nerve to attack a fire-spitting mountain like *Bismarck*.' *Bismarck* had opened fierce and accurate fire. Esmonde's Swordfish was hit on its starboard aileron four miles out. Tempering valour with sensible discretion, Esmonde decided to attack from where he was, off *Bismarck*'s port beam. Another 'vic' was led by Lt P.D. Gick, a most experienced pilot who had been an instructor at the Crail torpedo school, who decided that their first approach was not good enough. Still under fire he led his flight away from *Bismarck* and then back again, to drop on the port side.

'There was a *hell* of a sea running,' said Gick in a tape-recorded account after the war, 'and I really got down on the surface and kept losing sight of her behind the waves. On my left Pat Jackson [Sub-Lt (A) D.P.B. Jackson] was doing the same but on my right, Bill Garthwaite [Lt (A) W.F.C. Garthwaite, RNVR] a very recently qualified student of mine, who throughout his distinguished career in the Fleet Air Arm during the war seemed to be quite convinced that nobody could ever shoot him down, went in quite steadily and, when we were eventually sighted, drew all fire. It was his observer Anthony Gillingham [Sub-Lt W.A. Gillingham, RNVR] who commented afterwards. "He didn't mind the stuff going above him and below him but it was the beastly little balls of fire that nipped in between the wings that upset him."

'I like to think that I got to the right range. But she was a hell of a big ship and I have a horrid feeling I may have dropped a little too early. Anyhow I did and as I dropped, Bill and Pat also did so and we pulled up and turned away. By this time there was a hell of a lot of fireworks going on but my dear observer [Sub-Lt V.A. Norfolk, RNVR] had brought with him a very expensive camera and was determined to get some pictures and insisted that I should turn this way and that whilst he photographed the ship. Frankly, I thought the whole thing terribly unhealthy and stupid and in fact it very nearly was because when eventually he was satisfied and I got down on the water and flew away,

Bismarck did what I regarded as one of the most unsporting things of all which was to lob 16-inch shells at us, and in fact this caused great inconvenience because one of them landed ahead of us and I flew through the splash. The whole aircraft gained about thirty feet in altitude, and the only comment that came out of Sayer [the TAG, PO (Air) L.D. Sayer] during the whole four hours' flight was made then, when he said "Goddamn, some rotten sod's knocked the bottom out of my house". In fact, the splash had literally ripped the fabric off the bottom of his part of the cockpit and there he was, sitting gazing into fresh air for the rest of the trip home.'

The third flight, led by Lt H.C.P. Pollard, lost one of its members who strayed en route, but the remaining two pressed on and Swordfish 5K, piloted by Sub-Lt R.G. Lawson, RNVR, crossed *Bismarck*'s bows and attacked from the starboard side. It was his torpedo which one of the shadowing Fulmars reported causing a 'great black column of dense smoke rising from the starboard side'. Some of the other Swordfish sprayed *Bismarck*'s upper deck with machine gun fire as they broke off. *Bismarck*'s own flak was surprisingly ineffective, possibly because the ship was heeling under large amounts of wheel, and possibly because the gunners constantly over-estimated the Swordfish's speed.

It now only remained for them all to get back. It was nearly 2 a.m. and almost dark before the first Swordfish came in sight of *Victorious*—to the huge relief of Captain Bovell, who did his best to assist his straying flock by shining his largest signal projector in their direction and switching on all the flight deck lighting. When Rear Admiral Curteis, thinking of the U-boat peril, ordered him to switch the lights off, Bovell then began a very long flashing signal with his brightest lamp. All the Swordfish landed on safely, but two of the shadowing Fulmars were lost.

The pilot of one Fulmar, Lt Frank Furlong, RNVR, who had ridden Reynoldstown to his first Grand National Victory in 1935, and his observer, Lt A. John Hoare, had what Hoare called 'a quite remarkably improbable escape'. Hoare wrote in *Tumult in the Clouds*: 'We were flying at between 100 and 200 feet and a little south of west, on the most southerly leg of the search, in and out of snowstorms, sometimes skirting a particularly heavy one so as to be able to look. Our outward leg was to be two hours and almost at our turning point we both thought we saw the *Bismarck* and turned towards her into a snowstorm. When we cleared the snow there was no trace. Hallucination? Another ship? We could not radio a report on a glimpse from the corner of the eye and thereby send the fleet on a wild goose chase. We decided to make a small square search and turn for home. We found nothing and the light was going. A couple of hours later we knew we were in trouble. The fleet had altered course and *Victorious*'s homing beacon (as we later heard) was off the air. The official endurance of the Fulmar was some four and a half hours. We started a square search. (A course mate of mine, the observer of *Kenya*'s Walrus, told me later we had been heard and a searchlight was pointed skywards for us, but it was dark by then and the weather as thick as mud.) Frank, after nearly five hours, said: "We'd better come down while we still have an engine."

'We checked our ditching drill—and what a long hour that had been—and slowly motored into the wind. I fired Verey lights to give Frank a glimpse of where the water was. "Open everything that opens", said Frank, "and when we come down go straight to the dinghy. Stand by."

'There was a bit of a thump, a long pause when nothing happened, then a real thump and we were down.' But in spite of what Hoare calculated as odds of millions to one against being picked up in that remote area of the Atlantic, their dinghy was seen by the ss *Ravenhill* and they were both saved (only for Furlong to be killed later in the war).

After Gick had landed on and taxied up to the forward end of the flight deck, 'the bows dipped and a dirty great sea came over and salt water flew all over the aircraft. Sayer spoke again. "Ah well," he said, "that bloody hole's come in handy."'

Lawson's torpedo hit did not slow *Bismarck* down appreciably either and at 3 a.m. shortly after the Swordfish strike, luck turned *Bismarck*'s way. Perhaps *Suffolk* had grown somewhat overconfident after shadowing so successfully for so long, but when she steered back from the outward leg of a zigzag, *Bismarck* was gone. Lutjens had chosen his moment to alter to the west and increase speed. *Bismarck* was lost, and stayed lost for all that day and through the next night.

She was not found again until 10.30 a.m. on 26 May, when she was sighted by a Coastal Command Catalina flying boat, piloted by Flying Officer D.A. Briggs of 209 Squadron. *Bismarck* was then 700 miles west-north-west of Brest and as aggressive as ever. When Briggs inadvertently flew too close, *Bismarck*'s flak shrapnel fragments actually hit his Catalina's hull.

Briggs did not sight *Bismarck* again after breaking away under her attack, and the position he signalled was some twenty-five miles in error, but Force H, including *Ark Royal*, had been steaming up from Gibraltar during the night and at 11.14 one of *Ark*'s Swordfish, piloted by Sub-Lt J.V. Hartley sighted a German warship, and reported it as a cruiser. Seven minutes later, a second Swordfish piloted by Lt J.R. Callander confirmed that the warship was *Bismarck*.

She was making some 20 knots, going so well and already so far south that by morning of the next day, the 27th, she would be under the cover of the Luftwaffe, flying from France. Once again, the only hope of slowing her down lay in a torpedo strike. *King George V* and *Rodney* were still 135 miles to the north and *King George V* was short of fuel and could not increase speed. Fuel considerations lay heavily on the Commander-in-Chief's mind at every point in the *Bismarck* chase; the lack of a fast tanker to accompany the capital ships nearly allowed *Bismarck* to get away.

But for *Ark*'s aircraft, she certainly would have got away. Swordfish continued to shadow *Bismarck* in pairs for the rest of the forenoon. *Bismarck*'s gunnery was still excellent and, firing by radar, several times disconcerted the Swordfish with accurate flak even when they thought they were safely hidden in cloud. The weather was getting worse. Admiral Somerville was afraid of losing touch and so at 1.15 p.m., he ordered *Sheffield* to close the enemy and shadow from astern. This signal was received in *Ark Royal*, but not decoded in time to inform the strike of fifteen Swordfish, led by Lt-Cdr J.A. Stewart-Moore, which took off at 2.50, in very bad weather (*Ark*'s flight deck pitching up and down by as much as fifty feet).

At 3.20 the Swordfish picked up the radar contact of a surface ship. It was closer and more to the west than, strictly speaking, they would have expected *Bismarck* to be but nevertheless it must be *Bismarck*—who else could it be? The Swordfish dived on their attacking runs, descended out of cloud, and there was the target. Eleven of them had released their torpedoes before they realised the ship was actually *Sheffield*, who forebore to open fire, increased speed and evaded all the torpedoes. Two exploded when they hit the water and another three in *Sheffield*'s wake. They were all fitted with magnetic firing pistols which were clearly unreliable.

Shame-faced and chastened, the aircrews returned to *Ark Royal*. Apart from the jolt to their professional pride, they had jeopardised a friendly ship and, what was worse, allowed *Bismarck* to get that much further south. But Captain Maund soothed their ruffled feelings, told them to rest and get a meal, and signalled the results to the Commander-in-Chief (but prudently deciding that this was not the time to tell Admiral Tovey about the attack on *Sheffield*).

However, valuable lessons had been learned and when the second strike was launched, the fifteen Swordfish (many of them the same aircraft from the first strike) each carried a torpedo fitted with a contact firing pistol. This time the strike, led by Lt-Cdr T.P. Coode, was ordered to rendezvous with *Sheffield* twelve miles astern of *Bismarck* and take final departure from her.

The Swordfish, from 810, 818 and 820 Squadrons, flew off at 7.15, about thirty-eight miles north of *Bismarck*. Meanwhile *Ark Royal*'s own marvellous luck still held. While flying off aircraft she steamed at a steady speed across the periscope sights of U-556 which, however, had already expended all her torpedoes in a convoy attack.

From *Sheffield*, Coode learned that *Bismarck* bore 118°, range twelve miles. Conditions were stormy, with seven-tenths cloud between 2000 and 5000 feet. Coode had intended the strike, which was divided into six sub-flights, to make a co-ordinated attack from several sectors, to confuse *Bismarck*'s fire control, split up her AA defence and make it harder for her to evade the torpedoes. But in those conditions, the sub-flights became separated and in the end they all had to make their own way to the target and do the best they could.

Coode himself picked up *Bismarck* by radar twelve minutes after leaving *Sheffield*. Visibility had dropped to a few yards but his sub-flight dived from astern of *Bismarck*. Still blind in cloud, Coode watched the figures on his altimeter and when he reached 2000 feet began to worry about his height. At 1500 feet he wondered whether to continue the dive. At 1000 feet he felt sure something was wrong, but still the aircraft was completely enclosed in cloud. Coode held formation and at only 700 feet they broke cloud, just when they were nearly running out of height.

The first attack developed at 8.57 when Coode, deciding a slow beat up from astern against the wind would be suicidal, flew back into cloud and attacked again on *Bismarck*'s port beam. His sub-flight probably scored the torpedo hit aft which *Bismarck* signalled at 9.05. The second sub-flight, led by Lt D.F. Godfrey-Fausett, attacked from the starboard side. One of their flight, piloted by Sub-Lt D. Beale of 810 Squadron, became detached, flew back to *Sheffield* to get its bearings and returned again to attack. Beale's observer, Sub-Lt C.E. Friend had a close view of *Bismarck*. 'She impressed me at first by her great size,' he wrote in an account of the attack for his old school. 'She looked bigger than any warship I had seen before, as indeed she was. She had a wicked look because unlike British battleships of the time, which had plenty of space amidships between funnels and bridge, all her upperworks were together, giving her a humped appearance. As we closed her in a medium dive I could see, too, that she had an enormous number of close-range guns. She was in a wide slow turn.

'At 50 feet above the mountainous waves, Beale levelled out, just on her port bow, and about 1000 yards from her. He flew in towards her, straight and level, and sighting carefully. At 800 yards, the best range, he dropped the torpedo.

'None of us saw the torpedo running because he immediately threw the Swordfish on to its side in a fast turn away. At that instant *Bismarck*'s decks seemed to explode into crackling flame as she opened fire on us. The sea around was lashed by various sizes of shot and fragments and explosions were filling the air around us. Petty Officer Pimlott, the air gunner, in spite of the bumping jerkiness of Beale's jinking flight, was firing his gun in reply although it could have had little effect at that range other than to boost our morale.

'I was watching for any sign of a hit, and I was rewarded by the sight of a tall plume of water right amidships. Our torpedo had hit the target. Pimlott was doing a small jig as I excitedly told Beale. By turning the Swordfish quickly he too was able to see the splash subsiding. Thus all three of us saw our hit. We were now out of range of *Bismarck*'s close range weapons but her 5.25-inch guns were firing at *Sheffield*, who was rapidly coming into sight ahead of us. We flew close down her side making the usual "thumbs-up" sign of success, before setting off for *Ark Royal*.'

The last attack ended at about 9.40. No Swordfish were lost although several were damaged; 4C piloted by Sub-Lt F.A. Swanton had 175 shrapnel holes. Swanton and his TAG Leading Airman J.R. Seager were both wounded.

The 'slow turn' Friend had noticed was an involuntary one. By an extraordinary chance, the torpedo had hit aft in the one vital place that could halt *Bismarck*'s flight, by damaging her propellor shafts and jamming her rudder. Soon, to the astonishment of the shadowers, *Bismarck* was seen to be steering northwards, *towards* her pursuers, and shortly completed a full circle. Harried by torpedo attacks by the 4th Destroyer Flotilla through that night, *Bismarck* came under the heavy guns of *King George V* and *Rodney* soon after 10 a.m. the next morning. Stopped and battered into a blazing hulk, *Bismarck* still did not sink, and had finally to be dispatched by torpedoes from the cruiser *Dorsetshire*. A third strike of twenty-one Swordfish from *Ark Royal* was in the air at the time, but they were not needed, jettisoned their torpedoes and landed on.

Loss of *Ark Royal*

At 3.41 p.m. on 13 November 1941, *Ark Royal*'s charmed life came to an end. She had just taken part with *Argus* in yet another ferrying operation to Malta and was returning to Gibraltar when she was hit amidships on the starboard side by one of a salvo of three torpedoes fired by U-81 (Lt-Cdr Guggenberger). She took an immediate list, but still had steam pressure and was able to maintain power for lights and pumping. Unfortunately, the list was not corrected by counter-flooding sufficiently or in time. The ship continued to sink lower in the water. Eventually, the water level blocked the funnel uptake so that the boiler gases could not escape, and the one boiler room still steaming, on the port side, had to be shut down. All power was thus lost and pumping stopped. Despite the efforts of a tug, accompanying destroyers, and her own ship's company to save her, *Ark Royal*'s list steadily increased. The sea reached compartments previously unflooded. At 6.13 a.m. on the 14th, when she was only twenty-five miles from Gibraltar, *Ark Royal* finally turned right over and sank. All her people except one were saved.

Ark Royal's loss was a severe professional set-back for the Navy and a tremendous knock to the nation's morale. She was possibly the most famous ship on the Allied side. Her name was constantly in the news. Lord Haw Haw had claimed her sunk so many times, but she had seemed unsinkable. Now, she really was gone. It was like the end of an era.

Kirkenes and Petsamo

When in A.P. Herbert's words, Hitler 'leaped upon his largest friend' and attacked Russia in June 1941, the focus of the Navy's attention shifted from the North Sea to the Arctic. The rigours of the Arctic convoys were still in the future but in July the Russians had already asked for Allied assistance to stop the traffic of enemy shipping from the two northern ports of Kirkenes, and Petsamo, formerly in Finland.

It was politically necessary to give Russia all the assistance available as soon as possible and so on 23 July, *Victorious* and *Furious*, escorted by the cruisers *Devonshire* and *Suffolk* and six destroyers, sailed to carry out strikes with Albacores on the two ports. *Victorious* had twenty-one Albacores of 827 and 828, with twelve Fulmars of 809; *Furious* had nine Swordfish of 812 and nine Albacores of 817, nine Fulmars of 800, and four Sea Hurricanes of 880A Flight.

While the Albacores and Swordfish bombed or torpedoed the shipping which was reported to be 'massed' in the harbours, a top cover defensive screen of Fulmars would deal with the German fighter opposition, which was believed to be slight or non-existent.

That was the theory. The practice was somewhat different. A daylight strike by Albacores on any target, however lightly defended, was always fraught with hazards, indeed Admiral Tovey had expressed his misgivings about this operation. Worse, the force were sighted by a German shadower on the afternoon of 30 July, about eighty miles north-east of Kirkenes, as the carriers were preparing to launch aircraft, and all surprise was lost. Not for the last time in northern waters, politics outweighed tactics, and the strike was ordered to proceed. The Albacores took off in bright Arctic sunshine, with splendid visibility all round, to attack an enemy who turned out to be ready and waiting.

One of *Victorious*'s TAGs who survived the raid wrote a description of the debacle in 'Albacores in Action': 'Approaching the narrow entrance to a fjord, at cliff-top height, what seemed like thousands of guns opened up, the guns trained horizontally on us as we went by. I felt like throwing my Vickers "K" at them. We climbed to get over the hill at the end of the fjord and down the other side was the harbour, with about four small ships in it (massed shipping?). The ground gunners are still firing, and as we turn to starboard to line up for a torpedo run on the largest ship, I see a nice formation of Me110s half a mile away, at about 500 feet (of course they didn't know we were coming—they just happened to be there!). It's every man for himself. Torpedoes careering through the water, and Albacores going all ways, hotly pursued by Me110s. My aircraft did a smart about turn and retired at full speed (100 m.p.h.) the way we had come, followed by the rest of the sub-flight, hopping over the hill and down the fjord. My No. 3 hit the sea in flames; we cleared the mouth of the fjord and my No. 2 turned away towards the Russian coast, chased by two Me110s, which eventually shot him down. By this time we were about eight miles out to sea, and the last of the attacking Me 110s returned to base.

'Back over the carrier, and one aircraft is in a big heap on the deck; we landed, after it had been cleared, and then waited for the rest of the twenty aircraft that had taken off. We waited in vain. The score turned out to be eleven aircraft lost, and one TAG dead on arrival back on board.'

Two Fulmars were lost, having been generally out-performed by the opposing Me109s and 110s, and eight Albacores damaged. The Germans also used Junkers Ju87 Stuka dive-bombers in a combat role; one of them was remarkably shot down by the front gun of an Albacore flown by Lt J.N. Ball (who landed on *Victorious* with the only undamaged Albacore of the raid). The Fulmars claimed one Me109 and one Me110.

Victorious's Albacores had fired torpedoes at the gunnery training ship *Bremse* and four smaller ships. At Petsamo, *Furious*'s strike of twelve Albacores, six Swordfish with bombs and six Fulmars found no ships at all and slight fighter opposition. They lost three aircraft and four damaged. For *Furious* this was a disappointment which might have been worse. For *Victorious*, still a new carrier with an air group only just fully worked up, the Kirkenes episode was a disaster. Five officers and four ratings were killed, and another twenty officers and seven ratings became prisoners of war.

On the credit side, the chance had been taken during the operation to run the mine-layer *Adventure* to Archangel with a large cargo of mines, of which the Russians were said to be 'most appreciative'. But otherwise, Admiral Tovey's report was remarkable for what it left unsaid about this politically inspired operation and, after praising the ships' companies and aircrews who took part, the Admiral concluded somewhat bitterly that he trusted 'that the encouragement to the morale of our Allies was proportionately great.'

4

Worldwide Actions in 1942

The Gallant Sortie

One of *Ark*'s survivors was Eugene Esmonde, who had been transferred to her with his squadron after the *Bismarck* chase and, for the second time in the war, got away unscathed from a stricken carrier. On 11 February 1942, Esmonde went to Buckingham Palace to receive from King George VI the DSO he won for his part in the *Bismarck* episode. The next day, Esmonde went down to Manston, in Kent, where 825 Squadron, now reduced to six Swordfish, were stationed in case of a break-out of the German capital ships from Brest.

Scharnhorst and *Gneisenau* had arrived at Brest on 22 March 1941, after a sortie in the Atlantic in which they sank 115,600 tons of Allied shipping. They were joined there on 1 June by *Prinz Eugen* after her sortie with *Bismarck*. At Brest, all three ships were blockaded by sea and air and were primary torpedo and bomb targets for Bomber and Coastal Commands. All three were hit by bombs at some period during 1941, and in April *Gneisenau* was hit by one torpedo dropped by a Beaufort of Coastal Command (for which the pilot Flt Lt Campbell won a posthumous VC).

By the end of 1941 Hitler, obsessed with the notion that the Allies were preparing to invade Norway, and utterly confident of his own infallibility and flair for strategic decisions, decided that the three ships must return to Germany. He also decided that the shortest way was the best, through the English Channel. Hitler therefore gave the *Kriegsmarine* a flat ultimatum: either the ships returned to Germany, or they would be paid off in Brest and reduced to hulks, their crews being drafted into the Luftwaffe.

The break-out was meticulously prepared. The route was swept of mines, and special marker buoys and guide-ships were stationed. The heavy ships were to be escorted by five large and five light destroyers, being joined by E-boats at times en route. In a rare bout of co-operation with the German Navy, the Luftwaffe provided about 280 fighters, Me109s, Me100s, and the new Focke-Wulf Fw190s, to provide air cover of at least sixteen fighters over the force at all times. For the passage through the Straits, this number would be increased to thirty-two fighters. Such intense air and sea activity would certainly show on Allied radar screens but the Germans skilfully used jamming and electronic counter-measures against the radar stations along the south coast.

By January 1942 intelligence from various sources confirmed that a break-out was imminent. The Admiralty's forecast of German intentions was astoundingly accurate as to their strength, probable route to Germany and their air cover. Only the time of day was wrongly estimated. Nobody believed that Vice Admiral Ciliax, the German fleet commander, would dare to lead his ships through the Straits of Dover in broad daylight. From 1 February, reconnaissance patrols were stepped up and contingency plans brought forward to deal with the break-out.

Unfortunately, things went wrong for the Allies from the very first. The patrolling submarine *Sealion* missed the German ships as they left Brest roads at 10.45 p.m. on 11 February, and, by faulty radar and some bad luck, Hudsons of Coastal Command also missed them. Spitfires actually sighted the ships the next morning, but did not report until they flew back to base and it was well after 11 a.m. before the Flag Officer Dover, Admiral Ramsay, and the Admiralty had definite news that the German ships were at sea and in fact already very close to the Straits of Dover. This long delay in detecting the enemy caused every attempt to stop them to be hastily mounted and unconnected with other attempts. Small forces, of MTBs, bombers, and destroyers, attacked as they became available, in order to inflict at least some damage on the enemy ships before they steamed out of range. Of these attempts, the bravest, and probably the most hopeless of all, was the gallant sortie of 825 Squadron.

It was decided that Esmonde's squadron should attack at 12.45 p.m., which left very little time for briefing the fighter pilots of Esmonde's escort or for bringing more fighter squadrons to Manston. Esmonde knew what he faced, with or without an escort. Wing Commander Tom Gleave, the station commander, wished him luck before take-off and could not avoid noticing that 'although his mouth twitched automatically into the semblance of a grin and his arm lifted in a vague salute, he barely recognised me. He knew what he was going into. But it was his duty. His face was tense and white. It was the face of a man already dead. It shocked me as nothing has ever done since,' he said, in John Deane Potter's *Fiasco*.

Esmonde was supposed to have been escorted by five fighter squadrons, but when the first Spitfires joined him at 12.28 as he circled over Manston he decided to go ahead without waiting for the rest. He knew that, with a loaded Swordfish top speed of just under 90 knots, unless he hurried he would never catch up his targets. Two fighter squadrons, having missed Esmonde over Manston, flew over the Channel and engaged German fighters as Esmonde was making his attack. Two more squadrons searched for the enemy off Calais but failed to find them.

Shortly before Esmonde sighted the German ships, five MTBS from Dover made a torpedo attack but fierce gun and cannon fire from German fighters forced them to fire their torpedoes at ranges of about two miles and all missed. The MTBS, however, were then in the right place to pick up Esmonde's survivors.

The six Swordfish of 825 Squadron, flying in two flights of three aircraft each, reached the enemy about ten miles north of Calais. Esmonde's aircraft was last seen flying over the German destroyer screen in a hail of fire. Squadron Leader Brian Kingcombe, leading 72 Squadron of Spitfires, saw him go. His account is also in *Fiasco*: 'I went down to 100 feet, clipping the bottom of the clouds, and we managed to keep most of the German fighters off them. The Germans were firing heavy guns which threw up great mountains of spray like water spouts. The Swordfish flew straight into them. Mostly they were caught by *Prinz Eugen*'s flak and I saw the leader and two others go into the drink. They caught fire and went diving in flames towards the water.'

Esmonde's Swordfish was almost certainly shot down before he could complete his attack. He and his crew, his observer, Lt W.H. Williams, and his TAG, Leading Airman W.J. Clinton, were all killed. The other two Swordfish in his flight pressed on, to launch their torpedoes against *Prinz Eugen*. One of the pilots, Sub-Lt C. Kingsmill, RNVR, recalled an odd optical illusion. 'The tracer came floating gently towards us and then *whizzed* past. There were more and more large splotches in the sea as aircraft and ships fired at us and their shells burst into the waves. We were really in it now. Suddenly, I felt a sharp pain in my shoulder and my foot went squelchy. Oddly enough I didn't feel any more pain and managed to keep control of the plane.'

The third Swordfish was piloted by Sub-Lt B. Rose, RNVR, who was also hit and badly wounded in the back but also managed to launch his torpedo. The Swordfish were harried all the way down their attacking runs by the German fighters who, once again, were disconcerted by the Swordfish's slowness. The Focke-wulfs lowered flaps and undercarriage to bring down their air speed but still overshot again and again and had to make their attacks from dead ahead or dead astern. The Swordfish TAGS, with great personal courage, then stood up in their cockpits and turned round, so that they could warn their pilots of an oncoming fighter. But Kingsmill and Rose's Swordfish soon followed Esmonde's into the sea. The second flight of three Swordfish, led by Lt J.C. Thompson were last seen heading for the enemy. All three were shot down and there were no survivors. None of 825's torpedoes scored a hit.

The MTBs picked up five men of Esmonde's flight: Rose, Kingsmill, and their observers, Sub-Lts E. Lee and R.M. Samples, and the only surviving rating, Kingsmill's TAG, Leading Airman Donald Bunce. Esmonde was awarded a posthumous Victoria Cross. The four Sub-Lts were awarded DSOs, and Bunce the Conspicuous Gallantry Medal. Esmonde's body was recovered from the river Medway on 29 April.

The German ships made steady progress towards home until 2.13 p.m., off the Schelde estuary, when *Scharnhorst* hit a mine and came to a dead stop. Ciliax decided to transfer with his Chief of Staff to a destroyer but soon she suffered damage from a premature explosion of one of her own shells and Ciliax had to move to a second destroyer. Whilst crossing in the motorboat, he had to sit and watch *Scharnhorst*, able to steam again, pass close to him at 25 knots. Then, to add injury to insult, he and his party were bombed by a Dornier 217.

Coastal Command had three squadrons of Beaufort torpedo-bombers spread about the country, in Scotland, Cornwall and at Thorney Island in Sussex. All three did their best in attacks that afternoon and evening but because of poor visibility, poor briefing (because of misguidedly tight security), radio misunderstandings and the lack of enough aircrews with experience of attacking ship targets, none of their attacks had any success and several Beauforts were lost. Bomber Command did no better, because low cloud prevented attacks being made at enough height to use armour-piercing bombs and the bombs which were dropped all missed. Bomber Command, too, lacked aircrews specially trained for attacks against shipping.

By that evening, the Germans had some reason for celebration. Excellent advance planning, good co-operation, high speed, worsening weather, efficient fighter cover and accurate flak, disjointed enemy opposition and a slice of luck, had all enabled the German squadron to survive attacks by MTBs, by destroyers who came out from Harwich during the afternoon, and by various types of aircraft armed with guns, bombs and torpedoes.

But, at 7.55 p.m., off Terschelling, *Gneisenau* also struck a mine. She too was soon able to steam again but about two hours later *Scharnhorst* struck another mine and this time she was seriously damaged. The mines were the RAF's last laugh, having been laid during the last fortnight, or even possibly that very day, in the enemy's probable track. *Scharnhorst* reached Wilhelmshaven, at slow speed, in the early hours of 13 February. *Gneisenau* and *Prinz Eugen* reached the mouth of the Elbe later that day.

It had been a glorious, if tragic, day for Esmonde and 825 Squadron and a very energetic day for Fighter Command who had done well to escort our own aircraft, attack enemy aircraft and shipping in a confused tactical situation, in poor weather, and with very little up-to-date information. But nobody else emerged with very much credit. The realisation that heavy enemy warships had passed so close to the English coast for the first time since the French Admiral Tourville defeated an Anglo-Dutch force off the Isle of Wight in 1690 sent a spasm of shock and outrage through the whole nation. 'Vice Admiral Ciliax has succeeded where the Duke of Medina Sidonia failed,' thundered *The Times* on 14 February 1942, 'nothing more mortifying to the pride of sea-power has happened in home waters since the seventeenth century. . . .' The Prime Minister ordered an enquiry under Mr Justice Bucknill.

In the end, as sometimes happens, a tactical reverse proved to be a strategic gain. Those heavy ships were, of course, far less of a threat to Allied shipping at home in Germany than they had been at Brest. Their flight had shown that their position at Brest had become untenable, and Hitler's decision suggested that his thinking had turned on to the defensive. Hitler also felt that he had justified his opinions over those of his admirals, as he had so often done over his generals. He had always favoured the torpedo over the bomb as a weapon against shipping, but in 1938 his naval and air staffs had convinced him of the opposite, on probably the only occasion, he said with justifiable bitterness, 'they were ever in full agreement'.

As the Fleet Air Arm's first VC of the war, and indeed the first since the Arm gained its separate existence from the RAF, Esmonde passed into immortality. Admiral Ramsay signalled to the Admiralty: 'In my opinion the gallant sortie of these six Swordfish constitutes one of the finest exhibitions of self-sacrifice and devotion to duty that the war has yet witnessed.' One of the survivors was quoted in *Fleet Air Arm* as saying: 'I know that if Winkle Esmonde were alive and the same call came again, he would not hesitate to form another striking force, and we would follow him.'

As for the German ships, *Gneisenau* was hit by Bomber Command whilst in the floating dock at Kiel on the night of 26/27 February 1942. Her refit was eventually abandoned and she ended the war an unmanned, disarmed, melancholy hulk. *Scharnhorst*, always known as 'lucky *Scharnhorst*', went down under the guns of the Home Fleet off the North Cape on Boxing Day 1943. On 3 February, *Prinz Eugen* was hit by one torpedo from the submarine *Trident*, off Trondheim, which badly damaged her stern, but she survived the war to become one of the 'guinea pig' ships used in the atomic explosions at Bikini Atoll in 1948. Ciliax came out of the affair with less credit than almost anybody else. The German blue–jackets composed a comic song about him.

The New Boys

After more than two years of a hard-fought war in the air, there were not many survivors of those regular Navy 'straight ring' aircrew and the peace-time amateur flying enthusiasts who had started off on 3 September 1939. Whole training courses had been wiped out almost to the last man by accidents, on flight decks and in training flights, and in actions such as the disasters at Trondheim and Kirkenes. Those who survived were further depleted by the Navy's own insensitive treatment of aircrews. An air group normally joined an aircraft carrier for the whole commission, just like any other branch of her complement. But the air group was subjected to stresses and strains not experienced by the rest of the ship's company. The point was well put by Rear Admiral Jamieson, then *Ark Royal*'s engineer officer during the Norwegian campaign in his book on *Ark Royal*. 'The contrast between the risks run by aircrews and by the remainder of the ship's company is always a problem in a carrier. It is strange, and very humbling, for a ship's officer to sit at a table with a pilot or an observer who, between breakfast and dinner, has flown to the distant shore and been in action over enemy-occupied territory.'

In its ignorance, the Admiralty kept aircrews flying until they literally dropped or, in their own words became 'bushed or twitchy.' John Hoare, who made that miraculous escape during the *Bismarck* episode, wrote 'Later in the war when we were all that much more tired, I had officers under my command who should never again have been allowed near a carrier. I can see two of them now. One eventually blew his brains out with his service revolver and the other crashed on deck, but mercifully stepped out unhurt. As I had warned the Admiralty beforehand about sending them to sea, a little more attention might have been paid to Squadron Commanding Officers thereafter – one hoped.'

As the years passed, the rate of casualties amongst the older aircrews dropped, because there were so few of them left and they had grown very cunning, and because they were promoted to lead the training programme or became Commanders (Air) or Air Staff Officers of carriers. They were no longer as they called it, 'at the sharp end', flying in front-line squadrons. Fortunately for the Navy, and for the country, they had held the ring long enough while the new youngsters of the RNVR were taught their trade.

The youngsters were really *young*—all reminiscences of the Fleet Air Arm in the Second World War stress the extreme youth of the aircrews. They were in their late teens or early twenties. Twenty-five was ancient, while *thirty* was as old as Methuselah. Many of them joined in the middle of their apprenticeships or their college courses, some even straight from school, having opted to join the Navy under the 'Y' Scheme. Would-be pilots and observers joined at Lee-on-Solent, TAGS at HMS *Royal Arthur*, for four weeks' initial training, after which they all went to *St Vincent*, the training establishment at Gosport, for two months. They emerged as Acting Leading Naval Airmen, having done a course in boat work, signalling, ship's routines, the history of the Navy, knots and splices, ship and aircraft recognition, elementary navigation, and square bashing. The pilots then went to elementary flying training with the RAF and later to Kingston, Ontario, for service flying training. Later in the war, under a scheme inaugurated by Vice Admiral John Towers USN, a proportion of British pilots went to America and did their complete flying training at the great US air bases of Gross Isle, Pensacola and Miami.

Meanwhile, the observers did a gunnery course at Whale Island, two months' signalling and wireless telegraphy at HMS *Raven*, navigation and reconnaissance techniques at HMS *Condor* at Arbroath on the east coast of Scotland, and then to Greenwich, where pilots and observers, now Acting Sub-Lieutenants (A) RNVR, met again.

Greenwich was an oasis of civilisation in the rigours of aircrew training. There the Navy, evidently feeling that young aircrews were savage animals requiring a veneer of polish, provided gracious living, the Painted Hall, napkins, and the pursuit of Officer-Like Qualities (OLQS).

After Greenwich, the observers went to squadrons' working-up, while the pilots' paths split. Fighter pilots went to Naval Air Fighter School at HMS *Heron* at Yeovilton in Somerset, the TBR pilots to HMS *Jackdaw* at Crail in the Firth of Forth for torpedo training, the Walrus and ship's flight pilots to Lee-on-Solent. They did their Aerodrome Dummy Deck Landings (ADDLS) at HMS *Peewit*, in Angus, and joined working-up squadrons at Campbeltown (HMS *Landrail*) on the west coast of Scotland.

The TAGS, after *St Vincent*, went to Worthy Down (HMS *Kestrel*) in Hampshire for signals and wireless telegraph training and then to St Merryn (HMS *Vulture*) on the Cornish coast for air gunnery. Later in the war TAGS went to Canada for training, in Yarmouth, Nova Scotia.

The Indian Ocean

In the spring of 1942, the Admiralty diverted considerable carrier strength to the Indian Ocean for the first time, to meet a growing Japanese threat in the Far East, a theatre where the Allies had met almost constant defeat since Pearl Harbor. The two great capital ships of Force Z, *Prince of Wales* and *Repulse*, had been sunk by air attack in December 1941. British and Indian troops in Malaya had suffered a series of reverses in January 1942. Singapore fell on 15 February, Rangoon on 8 March. The Dutch government in Indonesia surrendered unconditionally on 9 March. At sea, in the air, and in the jungle, the Japanese seemed invincible.

In March, the remnants of the Eastern Fleet retreated to Ceylon where Admiral Sir James Somerville arrived on the 5th to take over as Commander-in-Chief. Somerville was as good as a tonic to a rather demoralised fleet. With all Force H's battle honours upon him, he had a splendid war record. He was a bold man and a witty man, outspoken, with an irreverent turn of phrase and a somewhat Rabelaisian sense of humour. When he first saw his fleet at sea, he signalled: 'So this is the Eastern Fleet. Well never mind. There's many a good tune played on an old fiddle.'

The Eastern Fleet was the largest which the Royal Navy had yet assembled in the war and it contained literally every ship the hard-pressed Admiralty could spare, having robbed other theatres to provide it. There were five battleships, three aircraft carriers, seven cruisers and fourteen destroyers. But four of the battleships, *Resolution, Ramillies, Royal Sovereign* and *Revenge* were old, slow, short endurance 'R' Class vessels dating from the First World War. The flagship *Warspite* was almost as old. One of the carriers was *Hermes* who had no fighters and only one squadron of Swordfish, 814.

The other two carriers were *Formidable*, recently returned after repair in the United States, and *Indomitable*, a new carrier of the *Illustrious* Class. *Formidable* had 820 Squadron, transferred from *Ark Royal* and re-equipped with twenty-one Albacores, and 888 Squadron of eighteen Martlets—the Fleet Air Arm name given to the American Grumman F4F Wildcat, a robust, serviceable single-seat fleet fighter, with a top speed of 330 m.p.h., an operational range of 850 miles, armed with four wing-mounted 0.50-inch machine-guns. *Indomitable* had twelve Fulmars of 800, nine Sea Hurricanes of 880, and two Albacore squadrons, 827 and 831, with twelve aircraft each. There were also 803 and 806 Squadrons of Fulmars ashore at Ratmalana, in Ceylon, and a handful of Swordfish of 788.

In all, the Navy could muster forty fighters and some fifty-five torpedo-bombers, but some of the aircrews lacked experience, and the Eastern Fleet generally was more formidable on paper than in fact. However, it was still a fleet in being, and as such a powerful threat to the flank of the Japanese advances in Burma. On 28 March, Allied intelligence reported a Japanese raiding force entering the Indian Ocean, steaming westwards, towards Ceylon.

To meet this threat, Somerville assembled his fleet at sea on 31 March, and deployed his ships in two tactical units: a fast Force A, including the two fleet carriers, under his own command, and a slow Force B, including the 'R' Class battleships under his second-in-command, Vice Admiral Algernon Willis. Somerville hoped to engage the enemy with Force A, using an Albacore strike by night as the main weapon, while Force B was in support.

In the event, the short endurance of many of the ships forced the fleet to return to refuel at the fleet base, at Addu Atoll in the Maldive Islands, before the Japanese could be brought to action. No fleet engagement in fact took place. Somerville apparently believed that a great chance had been lost. 'I fear they have taken fright,' he wrote, 'which is a pity because if I could have given them a good crack now it would have been timely.'

It might have been timely but it would have been most unlikely. Somerville might have written belligerently, but there is evidence that he was much more circumspect in the disposition of his ships and the direction of air searches. Some aircrews came to suspect that they were being sent to search in every direction where the enemy was *not*. It now seems extremely fortunate that no meeting with the Japanese ships took place, for the raiders were actually Vice Admiral Chuichi Nagumo's formidable First Striking Force, with five of the six carriers whose aircraft had devastated the US Pacific Fleet at Pearl Harbour and pounded the north Australian port of Darwin in February, four fast battleships, two cruisers, and escorting destroyers. Nagumo's air groups had more than 300 aircraft, most of them manned by pilots with flying experience going back to pre-war operations over China. Barring a miracle, any meeting with them would have resulted in yet another disaster for the Allies.

Nagumo failed to find Somerville's ships but snapped up the crumbs left for him. Whilst Vice Admiral Ozawa, with one carrier, cruisers and destroyers, raided the east coast of India, doing much as he pleased, Nagumo attacked Colombo with ninety-one bombers and thirty fighters on Easter Sunday, 5 April. The harbour had been cleared in time and the raiders did nothing like as much damage as they did at Darwin, but thirteen RAF Hurricanes were shot down, and a flight of six Swordfish of 814 from *Hermes*, on passage at the time from

Trincomalee to Minneriya. Seven Japanese aircraft were shot down.

The same day, Nagumo's dive-bombers surprised and sank the cruisers *Cornwall* and *Dorsetshire* which had been detached and on their way to rejoin the fleet. When the news of *Cornwall* and *Dorsetshire*'s loss reached the fleet, the aircrews were ready to go. 'Then why weren't we airborne and on our way? For Christ's sake why?' So wrote one of the pilots, Hugh Popham, in his *Sea Flight*.

'A last message from the two cruisers as they went down, and we raged and blasphemed with frustration. Commander Flying was bombarded with the demand, and could only answer: the Admiral says no.

'All day it was the same: the sour anger of enforced inaction, of a sapping impotence.'

One searching Albacore of *Indomitable*'s 827 Squadron sighted survivors in the water and reported them but this aircraft was lost. The report was picked up by a second Albacore, which, at the limit of its patrol, actually sighted the Japanese fleet. But the Albacore itself was sighted and attacked by a Mitsubishi A6M Zero, at that time the best fighter in the East. The Japanese pilot made several attacking runs which the Albacore, with a very experienced crew, managed to evade. The TAG, Leading Airman Gordon Dixon, was wounded on the first run but kept on firing until the Zero broke off. According to *Fleet Air Arm*, Dixon dug the bullet out of his arm with a screw-driver, put it in his pocket and said 'Must keep that as a souvenir.' He then fainted from loss of blood while the observer was rendering first aid. Years later, of the screw-driver anecdote, Dixon said 'All bull'.

For the next few days both sides searched for each other without success. On the 9th, Nagumo's carriers had refuelled and returned to strike at Trincomalee. They were sighted in time by a patrolling RAF Catalina and once again the harbour was cleared by the time the Japanese aircraft arrived just after dawn. The retiring strike aircraft sighted *Hermes*, one of the ships evacuated from Trincomalee, and came back three hours later when she was only sixty-five miles from Trincomalee, to inflict forty bomb hits in ten minutes, an onslaught which would have overwhelmed a much larger ship. *Hermes* turned over to port and sank in a few minutes with the loss of 304 lives. She herself had no fighters or aircraft of any kind embarked when she was attacked, but Hurricanes of 277 Squadron RAF and Fulmars of 803 Squadron arrived in time to chase away the last Japanese dive-bombers, who, however, had by that time also sunk the corvette *Hollyhock*, the destroyer *Vampire*, and two fleet auxiliaries. The whole day had been yet another smashing victory for the Japanese naval air arm.

After that, the Eastern Fleet's slow ships withdrew to Mombasa, while the faster ships went to Bombay. Somerville's fleet was still in being, but from now on, as he said himself, he would 'have to lie low in one sense but be pretty active in another—keep the old tarts out of the picture and roar about with the others.' But there was, in fact, little chance to roar about. By May 1942, the Eastern Army had reached the frontiers of Assam, after a retreat of 1000 miles, the longest in the history of British arms. The last American stronghold of Corregidor had fallen and the Japanese were masters of the Philippine archipelago. In six months they had won an empire which ran from Rabaul to Rangoon, and from Manchuria to Mandalay.

Yet, even in this time of tremendous triumph, the Japanese suffered their first serious reverse at the Battle of the Coral Sea. In the Indian Ocean, too, there was just the first faintest movement of the tide against them.

Diego Suarez

Seeing how spinelessly, without resistance or protest, the Vichy French had succumbed to Japanese pressure in Indo-China, the Allied Chiefs of Staffs laid plans to seize the Vichy island of Madagascar, which lay in an important strategic position athwart the route of the vital convoys up and down the east coast of Africa. No land-based air support was available, so the whole air effort for Operation IRONCLAD, as the Madagascar venture was code-named, was provided by the Fleet Air Arm from the two carriers *Illustrious* and *Indomitable*. *Formidable*, the third carrier in the Indian Ocean, remained with Admiral Somerville to give distant cover in case of a possible Japanese interference. However, as this operation coincided with the climactic struggle in the Coral Sea, the Japanese played no part.

Illustrious (Captain A.G. Talbot), newly repaired after her war damage, had two fighter squadrons, 881 with twelve Martlet Is, and 882 with eight Martlet IIs (with folding wings). *Illustrious*'s TBR squadrons were 801 and 829, equipped with twenty Swordfish.

Indomitable (Captain T.H. Troubridge), flying the flag of Rear Admiral D.W. Boyd, had three fighter squadrons, 800 and 806 each with twelve Fulmars, and 880 with nine Sea Hurricanes. Their TBR squadrons, 827 and 831, had a total of twenty-four Albacores. It was typical of the Fleet Air Arm, and the way in which its aircraft were gathered as they became available that these two carriers in company should be operating no less than five different types of aircraft; three British, one American, and one ex-RAF marque.

The first main landings took place on the western side of the isthmus of Diego Suarez before dawn on 5 May 1942, while the carrier air groups began a most enjoyable and successful two days of mixed air operations. *Indomitable*'s fighters strafed the airfield at Antsirane, whilst her Albacores bombed the hangars, destroying several Moranes inside. *Illustrious*'s Swordfish, with bombs and torpedoes, sank the Vichy submarine *Beveziers*, and the Armed Merchant Cruiser *Bougainville*, and drove the sloop, *D'Entrecasteux*, badly damaged, ashore. Both carriers provided reconnaissance, spotting and air strikes for the troops ashore, on demand.

The next day, the 6th, Albacores supported the advancing troops with bombing raids, and Swordfish carried out a simulated parachute drop to the south of Antsirane, as a decoy. One of 829's Swordfish, piloted by Sub-Lt (A) F.H. Alexander, RNVR, followed up their first success by sinking yet another Vichy submarine, *Le Heros*. *Illustrious*'s Martlets of 881 Squadron engaged and shot down four Morane-Saulnier MS 406Cs over the ship anchorage and two Potez 63.11s over the town. Meanwhile *Indomitable*'s Sea Hurricanes of 880 Squadron strafed the sloop *D'Entrecasteux*, whose guns were still giving trouble. After all the frustrations in the Indian Ocean of the previous month, this was an exhilarating experience, well described by Hugh Popham: 'We . . . flew in low all the way over the sea, over the low, bare hills of the northern part of the island, until the great harbour opened up in front of us, with the sloop in the shallows to the north of the town. She had steam up; and as we raced across the water, she let loose volumes of oily smoke that blew in a dense cloud towards us.

'One after another we tore into her, guns blazing. The tracer showed up briefly as it went sparkling into the smoke; our tracer going in and theirs coming out. I held it, thumb hard down and that intoxicating drumming of eight machine-guns shaking the whole aircraft, into the smoke until, a yard or two in front of me, the mast and aerials suddenly loomed up. I hauled back on the stick, and we went wheeling round for another run.

'None of us was hit; and when we left her, she was burning merrily.'

Vichy resistance in the north of the island ended on 7 May, by which time the carriers had flown 309 sorties and showed how far the Fleet Air Arm had come since Dakar, two years before. The two carriers withdrew for repair and maintenance, but *Illustrious* was back in September to cover the final operations in the island.

Over Madagascar, the carriers' aircraft exerted such pressure that the enemy believed that the attackers had no less than 150 aircraft available. For the Fleet Air Arm, the whole undertaking had been a very satisfying technical and professional success. 'We went into this operation without being fully worked up,' wrote Captain Talbot in *Fleet Air Arm*. 'We came out of it having had a real belly-full. From the air-operational, the deck-handling, the maintenance, in fact from every point of view, the experience was invaluable. Further, every flying crew was "blooded" although in most cases it was only by light anti-aircraft fire.'

The Desert Sailors

After *Formidable*'s departure in May 1941, no carrier operated regularly in the eastern Mediterranean, and after *Ark Royal*'s loss the following November, no carrier was permanently stationed in the Mediterranean at all. *Formidable*'s two fighter squadrons, 803 and 806, were sent ashore to Egypt where they joined 805, which had been operating in Crete, to form a large composite squadron, under Lt-Cdr A. Black, RN. This RN Fighter Squadron, as it was called, with 803 and 806 flying Hurricanes and 805 Martlets, joined a Fleet Requirements' Unit flying Fulmars, and three TBR squadrons, 815, 826 and 821, to become part of the Desert Air Force under the command of 201 Group RAF. This, after some controversy and argument between Cunningham, who naturally wanted the closest control of naval aircraft, and the Air Officer Commanding in Chief, Air Marshal Longmore, who wanted the same, became 201 (Naval Co-operation) Group in September 1941.

The Fleet Air Arm squadrons had not been trained or expected to operate in the desert but they took to their new task as to the manner born. The naval fighters carried out sweeps over enemy lines during Operation CRUSADER, strafing Afrika Korps armour and motor transport. They regularly provided fighter cover for coastal convoys supplying the Army, and over Tobruk during the siege of the city, and escorted daylight raids by RAF bombers. Though consistently outperformed by the Me109s they had to face (the Fulmars were especially out-classed and their pilots were constantly writing in their reports, 'Unable to close the range') they still had their success. Lt P.N. Chilton won the DFC, a very rare decoration for a naval officer, for shooting down three Ju87 Stukas in quick succession, but was then shot down himself in error by a supposedly friendly Curtis Tomahawk, happily surviving.

Another incident showed how death could overtake a pilot at the very moment of success. On 28 December 1941, Sub-Lt (A) A.R. Griffin, RNVR, flying a Martlet from an airfield near Tobruk, sighted four Italian torpedo-aircraft about to attack a convoy. He shot down one and forced two more into dropping their torpedoes wildly and at long range. The fourth pressed home its attack and was taken under fire by the convoy's guns. Over the R/T Griffin asked the ships to leave the target to him and the barrage ceased. The harassed Italian's torpedo missed, but Griffin's Martlet never pulled out of its attacking dive.

One TBR squadron, 815, under Lt Gick, who had been Esmonde's Senior Pilot, specialised in anti-shipping strikes, particularly in hunting for U-boats, using ASV radar and frequent ULTRA intercept information from Cairo. Their patrols were very long: six hours was normal, and Gick himself held the probable all-time Swordfish endurance record of eight and a quarter hours airborne— 'the next day, sitting down was not my favourite pastime,' he said. But they were successful. In nine months the squadron attacked over thirty U-boats, badly damaged three of them, and on 2 June 1942 shared in the sinking of U-652 with 203 Squadron.

The two Albacore Squadrons, 826 and 821, while keeping up their torpedoing expertise (826 attacked eight enemy ships in the last half of 1941) had another speciality—flare-dropping as path-finders for RAF Wellington night bombers. A pair of Albacores, their undersides painted matt black, engines in coarse pitch, would arrive over the target at six or seven thousand feet about fifteen minutes before the bombers.

Each with their thirty-two flares, the Albacores could illuminate the target for about an hour, and with practice, developed an excellent understanding with the RAF. A likely target, a gun position, a motor transport depot, or a troop encampment, would be reported by intelligence forward army patrols. An RAF reconnaissance aircraft, in one FAA pilot's words quoted by Terence Horsley in *Find, Fix and Strike*, 'would slip out while it was still daylight and get photographic confirmation. Before sunset we would be poring over the pictures, and at another aerodrome fifty miles away RAF Wellingtons would be bombing up. Then as soon as it was dark we'd take off loaded with flares, and a couple of 250-pounders for fun. Half and hour later we'd be over the target at 3000 feet, and if we looked upwards we could generally see the faint silhouettes of the Wellingtons a couple of thousand feet above us. We'd drop a flare and then, when we were sure, another and another, and the heavy stuff from the Wellingtons would come whistling down past us to explode with brilliant flames all round the target.'

Desert life had its own idiosyncratic flavour. It was all hard work and precious little play. Gick's observers, for example, with six hours or more in the cockpits every night, combining the duties of radar operator, navigator, rear gunner and telegraphist, could only keep going on Benzedrine, and a fortnight's operations at a time was about as much as they could take. For everybody, life was rough and tough at the Forward Landing Grounds, which was a polite name for stretches of desert cleared of rocks and other obstructions. The air and ground crews lived in ramshackle accommodation or in tents, on a diet of bully beef and hot tea. Water was often scarce, rationed down to half a gallon per man per day for all purposes. Transport was often commandeered from the enemy, as the battle ebbed to and fro. Names such as Maaten Bagush, El Adem, Sidi Barrani and Gambut found their way into squadron linebooks.

There were no landmarks in the desert, no trees, no shade, no women, nothing but work. In July 1941, the linebook of 821, commanded by a Royal Marine, Major Newsom, had a wry little verse:

> If on my Theme I rightly think
> There are five reasons why men drink
> Good wine; a friend; because I'm dry;
> Or should be by and by,
> And in an Albacore I fly.

In September 1942, the nadir of Allied fortunes in the desert, it seemed that even HMS *Grebe*, the headquarters of the Senior Officer, Naval Air Stations, at Dekheila, would have to be evacuated. But the crisis passed and Fleet Air Arm aircraft actually took part in the battle of El Alamein, flying thirty-seven sorties, night bombing and flare dropping over the battle front.

The TBR squadron's efforts in the desert were better appreciated by the RAF than by the Admiralty. While Air Marshal Tedder wrote of 'their magnificent work with and for the Wellingtons' and considered 'their night attacks were one of the decisive factors in crushing the enemy's attacks', the Admiralty were enquiring about what purpose the aircraft were fulfilling. In fact, they were keeping up a constant and telling pressure on the enemy. 821 Squadron alone flew 471 operational sorties, dropped 255 tons of bombs and over 2000 flares between June and November 1942.

The Malta Convoys

Shortly after the Diego Suarez operation, Admiral Somerville's fleet was further depleted, to meet the crisis of Malta. Since the departure of *Formidable* and the loss of *Ark Royal*, Admiral Cunningham and his staff had been wrestling with the virtually insoluble tactical problem of how to keep Malta properly supplied without enough naval air power at hand. Malta, strategically placed like the stopper in the neck of a vessel, was the key to the Mediterranean, and its fortunes were very closely linked with the progress of the land campaign in North

Africa—and vice versa. If Malta had sufficient sea and air power to disrupt Axis convoys, then the Allied cause prospered. Conversely, if the Afrika Korps pushed the Army back on land, then Malta suffered for it.

The New Year of 1942 had begun fairly promisingly. Rommel had been forced back to the borders of Tripolitania in December 1941 and airfields in Libya were again in Allied hands. Five supply ships reached Malta in January, escorted by Beaufighters of Nos. 252 and 272 Naval Co-operation Groups RAF. But Rommel also received reinforcements and on 21 January launched an offensive which recaptured the Libyan airfields and forced the 8th Army back to Gazala. Malta was furiously bombed again, from airfields in North Africa as well as Sicily. In February a brave attempt to pass three ships to Malta failed, none of them getting through. Worse still for Malta, a few days later the Axis passed a convoy successfully across to Tunisia, with strong battleship and cruiser cover, despite efforts to prevent it by aircraft from Malta and from No. 210 Group RAF in North Africa.

In March *Eagle* and *Argus* ferried forty-seven Spitfires to Malta from the west, and on the 20th a convoy of *Breconshire*, *Clan Campbell*, *Talabot* and *Pampas* sailed for Malta from Alexandria, escorted by four cruisers and seven destroyers under Rear Admiral Vian. Two days later, in the Gulf of Sirte, Vian's cruisers and destroyers fought a brilliant and very gallant action, repulsing a greatly superior Italian force which included the battleship *Littorio*. But the delay prevented the convoy reaching Malta before the next dawn. *Clan Campbell* was caught and sunk by bombers twenty miles out. *Breconshire* arrived, but was disabled and later sank. *Talabot* and *Pampas* reached Malta but were bombed and sunk alongside, on 26 March.

In April, the US carrier *Wasp* flew in another forty-seven Spitfires to Malta on the 20th, but the new arrivals were spotted on radar screens in Sicily and the Luftwaffe had destroyed them all within three days. On 9 May, *Eagle* and *Wasp* again, flew in sixty more Spitfires. But all these frantic efforts by sea and air were not nearly enough. Supplies of food, fuel and ammunition in the island continued to sink, seemingly inexorably. By June 1942 the situation of the people and garrison of Malta was desperate. Two more large convoys were sailed, one code-named HARPOON with six merchant ships from the west, and the other VIGOROUS with eleven merchant ships from the east. Both met tremendous Axis resistance. VIGOROUS had to turn back, with only thirty-three percent ammunition remaining, having experienced, in Vian's words, 'all known forms of attack', including the battleships *Vittorio Veneto* and *Littorio*, and returned to Alexandria. Two ships from the HARPOON convoy reached Malta with 15,000 priceless tons of supplies.

1 Off southern Ireland, 17 September 1939: *Courageous* sinking after hits by torpedoes from U–29, the first Royal Navy ship to be sunk by enemy action in the Second World War. Photo taken from the destroyer *Impulsive*.

2 Gloster Sea Gladiator, the Fleet Air Arm's only single-seater fighter at the outbreak of war, converted for fleet service by fitting catapult points and arrestor hook to the RAF version.

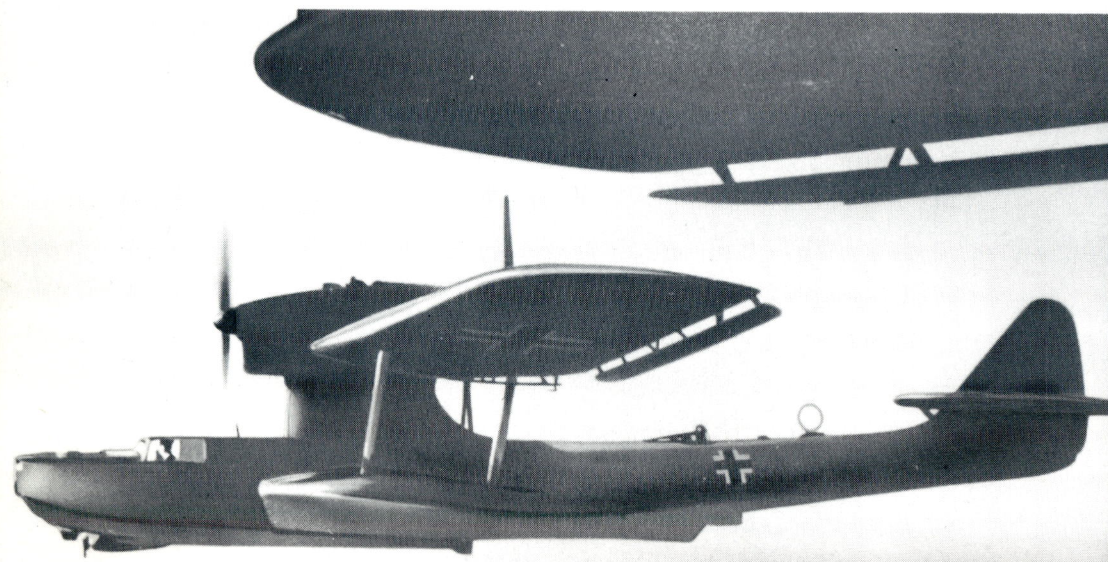

3 *top* A Blackburn Skua fighter-dive-bomber: though slower and clumsier than almost every opponent it ever flew against, it achieved success over Norway, largely due to the skill of its crews.

4 *bottom* Dornier Do.18 flying boat, of the type shot down by a flight of Skuas from *Ark Royal* in the North Sea on 25 September 1939, the first German aircraft of the war to be destroyed by British forces.

5 *Furious*, a much loved ship, converted from an 18-inch gun cruiser into an aircraft carrier in 1917.

6 Fairey Swordfish L9781, of 820 Squadron (the co's aircraft, No. 650), in flight over *Ark Royal*, c.1939/40.

7 *top* Force K in the south Atlantic, searching for *Graf Spee*, November 1939. A Skua has just been launched from *Ark Royal*'s port catapult with the battle-cruiser *Renown* refuelling a destroyer in the background.

8 *bottom* Skuas of 800 and 803 Squadrons on the airfield at Hatston in the Orkneys, from where they flew to sink the German light cruiser *Konigsberg*, 10 April 1940.

9 Lt Ronnie Hay, DSC, Royal Marines, who flew a Roc over Norway. 'Shouting "Fire" to the gunner, I heard one bang and silence. All the guns had jammed. . . .'

10 The Norwegian Sea, 4 p.m. on 8 June 1940. The German battle-cruiser *Gneisenau* opening fire on *Glorious* at a range of 28,000 yards. Photo taken from *Scharnhorst*.

11 Italian equivalent of 'Are We Downhearted?' Propaganda poster of 1940, showing a supposedly invulnerable Taranto.

12 *top* The Luftwaffe's revenge on *Illustrious*: a 500-kg bomb has pierced the flight deck and the explosion has blown out the after lift.

13 *bottom* *Illustrious* still under Luftwaffe attack whilst alongside Parlatorio Wharf, Malta, January 1941. The ship's flight deck can just be seen through the smoke to the right of the large crane.

14 Officers and men of 825 Squadron decorated for their part in the sinking of *Bismarck*: from left, Lt P.D. Gick, Senior Pilot (DSC), Lt-Cdr Eugene Esmonde, CO (DSO), Sub-Lt V.K. Norfolk, RNVR, Gick's Observer (DSC), Acting PO (Air) L.D. Sayer, Gick's TAG (DSM), and Acting Ldg Seaman (Air) A.L. Johnson, Esmonde's TAG (DSM). Photo taken in *Ark Royal*.

15 The powerful effect *Ark Royal* had on the German imagination: a painting of her flying off Swordfish against *Bismarck*, by Walter Zeeden (from *Das Geheimnis du Bismarck* by Fritz Otto Busch).

16 Some of *Ark*'s survivors safely ashore.

17 An Albacore taking off from *Victorious*.

18 Supermarine Walrus amphibian being launched from *Warspite*. *Illustrious* in the background.

19 Swordfish TAG, 1942.

20 *Hermes* sinking off the east coast of Ceylon, 10 a.m., 9 April 1942, after an attack by Japanese aircraft from Admiral Nagumo's carrier force and suffering forty bomb hits in ten minutes.

21 *Hermes*' flight deck, listing as the ship sank. Photo taken with a pocket camera by Ordinary Seaman (later Sub-Lt RNVR) Charles Morgan and successfully developed after having been in the Indian Ocean for four hours with its owner!

22 *Indomitable* in the Indian Ocean, 1942.

23 Early morning physical training on board *Victorious* at Scapa Flow, March 1942.

24 Three Albacores and a Swordfish on *Indomitable*, July 1942.

25 *Eagle*, hit by torpedoes from U–73 and sinking, 11 August 1942, while PEDESTAL convoy steams on.

26 Seafires of 885 Squadron on the deck of *Formidable* in Force H for Operation TORCH.

27 Lt B.H.C. Nation, Martlet pilot, who accepted the surrender of Blida airfield on behalf of the Allies on 8 November 1942.

28 Two escort carriers, *Biter* and *Avenger* (left), returning from TORCH. *Avenger* was torpedoed and sunk by U–155 near Gibraltar on 15 November 1942.

29 Seafire prang of 807 Squadron on *Battler* off Salerno, September 1943.

30 *Formidable* on her way to cover the landings at Salerno, September 1943. On deck, Seafire IIICs on out-riggers, and seven Martlet IVs of 888 and 893 Squadrons.

31 Fighter Catapult Ship *Springbank*, with Fairey Fulmar of 804 Squadron embarked.

By July, five merchant ships, three in March and two from VIGOROUS, had reached Malta that year, at the cost of one cruiser sunk and three damaged, an anti-aircraft cruiser damaged, eight destroyers sunk and five damaged, and seven merchantmen sunk and two damaged. The War Cabinet now decided that these attempts to relieve Malta were steadily bleeding away Allied strength without significantly helping Malta. It was therefore decided that a relief convoy for Malta should now be given priority over every other Allied naval commitment anywhere in the world.

In the early hours of 10 August 1942, providentially shrouded by thick fog from watchers on the Spanish shore, the great relief convoy, known as Operation PEDESTAL, passed eastwards through the Straits of Gibraltar. There were fourteen fast merchant ships, including the tanker *Ohio*, escorted by a powerful fleet: the battleships *Nelson* (flying the flag of Vice Admiral E.N. Syfret) and *Rodney*, the carriers *Indomitable*, *Victorious* and *Eagle*, with seven cruisers and twenty-four destroyers. In addition *Furious*, with an escort of eight destroyers, was to fly thirty-eight more Spitfires into Malta.

Aircraft carriers had covered Malta convoys before (*Ark Royal* the HALBERD convoy in September 1941, *Eagle* and *Argus* the eastbound HARPOON convoy in June) but PEDESTAL was the strongest force the Fleet Air Arm had ever fielded. Between them, the three carriers had seventy-four fighters. *Victorious* had disembarked six Albacores of 832, to make way for six Fulmars of 884 and six Sea Hurricanes of 885. These, with the twelve Fulmars of the 'resident' 809, gave her twenty-four fighters embarked, although the Hurricanes without folded wings, had to be stowed as a permanent deck-park, on outriggers projecting over the starboard side. *Indomitable* had twenty-four Sea Hurricanes of 800 and 880, and ten Martlets of 806. *Eagle* had twelve Sea Hurricanes of 801. *Indomitable* also had her Albacore squadrons, 827 and 831. *Furious* had six Sea Hurricanes of her own 804 Squadron.

PEDESTAL was one of the Fleet Air Arm's most spectacular set-pieces of the war. The theatre was the western Mediterranean. The aircrews had the eyes of the fleet and of the Merchant Navy on them, with visibility excellent, day after day. The objective, the prize, was the survival of Malta. On that first day, the mood in the carriers was, as one pilot in *Indomitable*'s 880 put it, 'cheerful, resolute, and as taut as wire.'

The tension was tightened another notch in the afternoon when two of *Victorious*'s Martlets were scrambled to intercept a shadower which, however, proved to be a solitary Vichy flying-boat, on its way to North Africa. It was ruthlessly, if distastefully, shot down into the sea. The alarm it would have raised might have made the difference between life and death to Malta.

It was in fact the enemy who struck first on the 11th, next day. Unseen by the Albacore A/S patrols and undetected by the destroyers, U-73 (Lt Rosenbaum) penetrated the screen and fired a salvo of four torpedoes at *Eagle*. The torpedo tracks were actually sighted by Sub-Lt Chancey Parker coming on to land on *Victorious* in his Fulmar. 'I looked over the side and saw three or four torpedo tracks very close together and heading straight for *Eagle*,' he wrote in *Send Her Victorious*, 'so I shouted to my telegraphist Leading Airman Ford to broadcast a warning, but they probably hit as the message was received.'

The four torpedoes hit *Eagle* port side, at 1.15 p.m. Hugh Popham was looking across, in the cockpit of his Hurricane on *Indomitable*'s flight deck, to see what the flight engineer was saying when, 'as I did so, I stared in shocked surprise beyond him to where *Eagle* was steaming level with us, half a mile away. For as I turned smoke and steam suddenly poured from her, she took on a heavy list to port, and the air shook with a series of muffled explosions....' Listing to port, she swung outwards in a slow, agonised circle, and in seven minutes turned abruptly over. For a few seconds longer her bottom remained visible; and then the trapped air in her hull escaped, and with a last gust of steam and bubbles she vanished. All that remained was the troubled water, a spreading stain of oil, and the clustered black dots of her ship's company.

One of the 'black dots' was Arthur Thorpe, the Exchange Telegraph correspondent (a survivor of *Ark Royal*, later killed on the Normandy beaches), who felt a shock at the base of his spine and writing in *The Daily Telegraph* of 14 August 1942, said he 'knew it was a depth-charge from a destroyer hunting the U-boat responsible. "She is going" gasped one of the men. Then came a mighty rumbling as the sea poured relentlessly into the *Eagle*, forcing out the air. The water threshed over her in a fury of white foam and then subsided. She had gone.' Destroyers picked up Captain L.D. Mackintosh and some 900 of *Eagle*'s company, but 250 men, several of her priceless fighters and the Italian ensign from Port Sudan, went down with her. *Eagle* had made nine aircraft ferry trips, and delivered 183 Spitfires to Malta in 1942. As the official naval historian remarked in *The War at Sea 1939–1945*, 'If we had to lose her it was appropriate that her grave should be in the Mediterranean, whose waters she had known so well.'

All day the fighter pilots waited for action; two in the air above the convoy, two on deck strapped in their cockpits and the engines warmed up, four at immediate notice in the ready-rooms. At last, towards dusk, the convoy came within range of Sicilian airfields and at 8.45 p.m. some thirty-six Ju88s and Heinkel IIIs attacked from the east but were driven off with the loss of a possible four aircraft. Popham was one of those who scrambled: 'The sight we saw took our breath away. The light was slowly dying, and the ships were no more than a pattern on the grey steel plate of the sea; but we had left them sailing peaceably through the sunset, now they were enclosed in a sparkling net of tracer and bursting shells, a mesh of fire. Every gun in fleet and convoy was firing, and the darkling air was laced with threads and beads of flame.'

Unfortunately, the convoy were firing at anything that flew and Popham, who was short of fuel, had an anxious time before he managed to put his Hurricane down on *Victorious*, although she was swinging hard at the time and tried to deter him by waving him off. He got out of his cockpit just before it went up in flames. Meanwhile, Captain Troubridge in *Indomitable* had decided that his carrier was useless without her fighters and decided to land them on willy-nilly. He steamed at 26 knots, on a straight course, showing deck lights, until eight of his straying sheep had returned. Even so, some were fired upon by 'friendly' guns, among them Lt-Cdr Judd, Commanding Officer of 880, who landed on, jumped out of his cockpit, ran across to the offending guns, seized the Lieutenant who was officer of the quarters by the throat and roared: 'You bloody useless bastard! You brainless oaf! Don't you know a Hurricane when you see one!'

There was no aircraft to take Popham back and he consequently missed an eventful day for *Indomitable*. The first attack, of about twenty aircraft, began at about 9.15 a.m. the next day, and attacks of one kind or another lasted for most of the day, growing in weight and frequency as the convoy steamed nearer and nearer to the Luftwaffe airfields in Sicily. Seventy aircraft took part in a prolonged attack from 12.15 to 1.45 p.m. which, however, seriously damaged only one ship, the merchantman *Deucalion* which was later sunk.

The attackers tried every kind of weapon: there were Stuka Ju87 dive-bombers, Ju88 high-level bombers, torpedo-bombers with a new aerial torpedo called a 'Motobomba FF', Heinkel IIIs laying mines in the convoy's path. There were some unusual variations, among them a radio-controlled Savoia loaded with explosive, whose crew bailed out before the bomber crashed (happily for the convoy) somewhere in North Africa. During a lull, two Reggiane 2001 fighters swooped low and flew along *Victorious*'s flight deck. They looked very like Hurricanes whose pilots had had a touch of the sun and, as somebody said at the time, 'they were silly twats, beating up the fleet at a time like this. Serve 'em right if somebody fired at them.' But nobody did, and they each dropped a bomb, one of which fell overboard and the other which hit the flight deck, broke up and did very little damage, very fortunately for *Victorious*.

During the afternoon the Italian submarine *Cobalto* was depth-charged to the surface and then rammed and sunk by the destroyer *Ithuriel*. In the evening, the air battle broke out again, with the heaviest raid of the day, and of the whole operation: over a hundred aircraft, Ju87s, Ju88s, Cant Z1007Bs, Savoia SM79s, Messerschmitt Me109Gs, Me110C-4s, Reggiane 2001s and Macchi C202s. The fleet fighters went up again to meet them, but this time, a formation of four Ju87s and eight Ju88s penetrated the AA barrage and fighter cover and attacked *Indomitable*. Watchers saw her disappear in the leaping columns of water, just as *Illustrious* had once done. *Indomitable* appeared out of the clouds of spray but one 1000-lb bomb had pierced the flight deck forward of the forward lift and exploded in the entrance to the hangar, killing or wounding many of the men working there, blowing the massive lift out of its mountings and leaving it cocked on a slant, projecting some two feet above the flight level. A second bomb pierced the armoured deck right aft, damaging the after lift and burning out officers' cabin flats. A third bomb hit the port side just above the water line and blew a thirty-foot hole in the ship's hull. It also wrecked the wardroom ante-room and killed a group of off-duty aircrew, including several Albacore pilots and observers, who were in the room at the time. The destroyer *Foresight* was torpedoed and sunk during this attack.

But the enemy did not escape unscathed. Exact figures are difficult to determine but the Axis lost between thirty and forty aircraft during the convoy's passage, the great majority of them to the guns of the fleet's fighters. Lt R.C. Cork, DSC, of 880 Squadron shot down five enemy aircraft personally, and later took over command of the squadron after his Commanding Officer, Judd, was shot down and killed. In all, thirteen aircraft were lost besides those that went down in *Eagle* or were damaged in *Indomitable*. Eleven aircrew were lost: an observer, two TAGS and eight pilots, including two squadron commanders. Lt-Cdr R.L. Johnston, Commanding Officer of 806, who was badly wounded in combat, managed to put his Martlet down on *Indomitable*'s deck, but the arrester hook broke out. The Martlet went over the side and sank, taking Johnston with it.

Half of *Indomitable*'s fighters were in the air when she was hit, and had therefore to land on *Victorious*, the only remaining deck. By the time they and *Victorious*'s own fighters had all landed on, the deck was choked back to the after deck barrier, preventing the ship launching aircraft if she were herself attacked. Any damaged fighters were therefore just run off the back of the round-down and dropped in the sea.

The convoy was now approaching the narrows of the Skerki Channel between North Africa and Sicily and at 7 p.m. *Victorious* and the other heavy ships hauled round to the west and departed, as previously planned. Command of the convoy devolved upon Rear Admiral Burrough in the cruiser *Nigeria*.

The day's adventures were by no means over. At dusk, the Italian U-boats *Dessie* and *Axum* attacked the convoy and *Axum* scored one prodigiously successful salvo of three torpedoes which (probably) hit *Nigeria*, which had to return to Gibraltar, the anti-aircraft cruiser *Cairo*, which had to be sunk, and the tanker *Ohio* which caught fire but steamed on. Admiral Burrough had to shift his flag to the destroyer *Ashanti* and, as he was doing so, the convoy was attacked again, by a force of some eighty bombers, flying in as the sun went down. The convoy had no fleet fighter cover and the fighters from Malta had just left. The ships were most unfortunately bunched up and, for the time being, at a disadvantage. The attackers got in amongst them and sank the merchantmen *Empire Home* and *Clan Ferguson; Brisbane Star* was also damaged but kept steaming. In the middle of this confusion, the Italian U-boat *Alagi* put one torpedo into the cruiser *Kenya*, and for some time the whole convoy was in a state of utter pandemonium, with a storm of gunflashes, bomb and torpedo explosions, streaks of flaming tracers threshing to and fro overhead and ships steering in all directions to avoid bombs, torpedoes and each other.

By midnight the convoy had restored itself to some appearance of order and was rounding Cape Bon when it was attacked by German and Italian fast E-boats which had been lying in wait close inshore. Their attack was the most effective of any. Between 3.15 and 4.30 their torpedoes sank the merchantmen *Wairangi*, *Almeria Lykes*, *Santa Elisa* and (probably) *Glenorchy*, and disabled the cruiser *Manchester*. She was scuttled next day—prematurely, as the subsequent court martial decided.

From dawn on the 13th, when the convoy was entering its last lap, the air attacks began again. The freighter *Waimarama*, carrying petrol and ammunition, was hit by one bomb and simply disintegrated in a sheet of flame, leaving some survivors swimming in the water. A Ju88 crashed on *Ohio*'s forecastle but she steamed on until incredibly, she was hit a third time, by a Stuka which landed on her poop. Her engines broke down and she came to a stop. Another merchantman, *Dorset*, was sunk that day and only three of the convoy, *Port Chalmers*, *Rochester Castle* and *Melbourne Star*, reached Malta that evening. *Brisbane Star* struggled in a day later.

Ohio was taken under tow and, amid great jubilation, reached Grand Harbour on the 15th with her 10,000 tons of petrol and kerosine. Her Captain, D.W. Mason, later received the George Cross.

From the PEDESTAL convoy, one aircraft carrier, two cruisers, a destroyer and nine merchantships had been lost, another aircraft carrier, two cruisers and three merchantmen had been damaged. Two Italian submarines were sunk. Nor did the convoy relieve Malta. Its arrival was only a reprieve and by December the island was once again in desperate straits and there was actually discussion of the date of Malta's surrender to the enemy. In November another convoy, STONEHENGE, unnoticed and unremembered, reached Malta, after the cruiser *Arethusa* had been seriously damaged and towed through a rising gale to Alexandria. Only in December, with the safe arrival of the PORTCULLIS convoy, was Malta finally saved. By that time the TORCH landings in North Africa had already begun to transform the strategic situation in the Mediterranean.

Operation TORCH

The landings in North Africa marked the end of the defensive phase for the Allies. Now, the war began to turn against the Axis. The German Sixth Army was held at Stalingrad. The Afrika Korps was decisively defeated and in retreat from El Alamein. In the Far East, the struggle on Guadalcanal was finely balanced, but in a few weeks would shift against the Japanese.

For the Fleet Air Arm, TORCH opened up a new concept of the war, a new world, in fact, where it was no longer a novelty but actually a commonplace to go on deck in the morning and see at least one and sometimes several carriers operating in company, where the deck parks were full of aircraft and many more were waiting ashore.

For TORCH, the fleet carriers *Formidable* and *Victorious* formed part of Force H, a strong group of carriers and capital ships to cover the landings from any possible interference by the Italian battle-fleet. There were no less than seven carriers for the operation as a whole, with thirty-seven Albacores and 137 fighters of various types embarked. *Formidable* had six Seafires of 885, six Martlets of 888, eighteen Martlets and Fulmars of 893, and their old familiars, 820 Squadron with twelve Albacores. *Victorious* had one of the last operational Fulmar squadrons, 809, under Captain R.C. Hay, Royal Marines, which was not embarked for fighter duties but had been specially trained for tactical reconnaissance at the Army Co-operation School. The fighter squadrons were twelve Martlet IVs of 882, and six Seafires of 884; one flight of 832 Albacores had been disembarked to Manston in Kent before sailing, so the ship had fifteen Albacores of 832 and 817.

While *Formidable* and *Victorious* held the distant ring with Force H, *Furious*, *Biter* and *Dasher* operated closer inshore with the Centre Naval Task Group, covering the landings at Oran. Similarly,

Argus and *Avenger* operated with the Eastern Naval Task Force off Algiers. Five American carriers led by *Ranger* covered the western landings in Morocco.

Furious had 807 Squadron, led by Lt-Cdr A.B. Fraser-Harris (who had flown Skuas in Norway), the first Seafire squadron to land on a carrier. This was the beginning of a prolonged love-hate relationship between the Fleet Air Arm and the Seafire. A standard Spitfire VB fitted with an arrester hook and catapult points carried out deck-landing trials in *Illustrious* late in 1941. This marque, simply renamed Seafire IB, had a top speed of around 350m.p.h., a range of 445 miles and was armed with a 20–mm Hispano cannon and two 0.303-inch Browning machine-guns in each wing. It could also carry one 500-lb bomb. The Seafire, as expected from its lineage and past RAF reputation, was a beautiful aircraft to fly and to fight. But it was somewhat short ranged for fleet work and far too fragile for the rough and tumble of deck-landing life. Its weak, narrow under-carriage, its long propellor and its excellent flying qualities—which made it tend to float onwards after the engine had been 'cut'—rendered it extraordinarily liable to all manner of deck-landing accidents, and most particularly to barrier entries.

For TORCH, *Furious* had twenty Seafires, of 801 and 807, with eight Albacores of 822. *Argus* also had Seafires: twelve of the re-equipped 880 from *Indomitable*. The three smaller escort carriers were wholly equipped with Sea Hurricanes—*Biter* had fifteen of 800, *Dasher* fifteen of 804 and 891, *Avenger* fifteen of 802 and 883—and represented a new departure in maritime air power. They were fighter assault carriers and were destined to play a very important part in amphibious operations in the Mediterranean, and later in the Far East. Their task was to operate close inshore, to provide constant fighter cover over the assault beaches and shipping during the assault phase, and afterwards, until airfields had been established ashore (although everybody hoped and trusted that this interim period would not last too long). Most of the British aircraft which took part in TORCH were painted with a plain white star instead of the usual roundel. This was partly to assist Americans ashore to identify Allied aircraft, but mainly to reassure the French; during TORCH, as at Mers el Kebir, Dakar and in every other operation in which the French nation was involved during the Second World War, the reaction of the French on the spot was impossible to predict.

D-Day was 8 November 1942, and the first strikes took off at dawn. The main target was a large salt-lake, about twelve miles long and a mile wide, which was some seven miles inland of the city of Oran. There, on the north-eastern corner of the lake south of Oran, was the large military airfield of La Senia. Tafaraoui, the civil airport, was five miles to the south. A force of Albacores, led by Lt (A) J.G.A. McI. Nares, escorted by Seafires and Hurricanes, attacked these two airfields. The Seafires met heavy and light calibre flak but flew down to strafe and silence the guns. (Fraser-Harris was hit, force-landed, set fire to his aircraft, taken prisoner, and released by an American armoured column two days later.) Meanwhile, the Seafires were engaged by some Dewoitine D520 fighters and Sub-Lt (A) G.C. Baldwin, DSC, shot down one of them—very probably the very first Seafire 'kill' of the war. Nares's Albacores dropped leaflets over the area and then bombed the hangars of La Senia, destroying no less than forty-seven Vichy aircraft in the hangars or dispersed around the field. Nares led through intense flak and Dewoitine attacks, and was himself shot down in flames. While the Hurricanes shot down five Dewoitines, four Albacores were lost, although the crews of three of them were saved.

At Algiers, four Martlets of 882 had taken off from *Victorious* at 5.45 a.m. to patrol over another large military airfield at Blida, thirty miles south-west of the city, their orders being to shoot up any Vichy aircraft which looked like taking off. They duly machine-gunned and deterred two such aircraft before returning to the carrier.

At 8 a.m. another section of four Martlets took off for the same purpose and destination, and thereby gave rise to one of those stories the participant could tell his grandchildren. After circling over Blida for half an hour, the Martlet section leader, Lt (A) B.H.C. Nation, saw people waving white handkerchiefs. Peasants and workers in the surrounding fields were also waving in an apparently welcoming manner. Nation reported this activity to *Victorious* but the staff on board doubted whether the wavers were as friendly as they seemed, and asked him to confirm that he was flying over the right airfield. Nation looked down at the name BLIDA painted in huge white letters on the ground and replied affirmative.

Strictly speaking, as the land campaign was under American command, a US officer should have accepted any surrender and Admiral Lyster in *Victorious* offered to fly his American liaison officer Captain Hanson, US Army, out to Blida. But Hanson generously said that the honour should go to the Fleet Air Arm. Nation was ordered to detail one pilot to land. Unsurprisingly, he detailed himself, telling his three pilots to patrol above.

Nation taxied his Martlet over to the hangars where he found quite a large reception committee of French officers who took him to the station commander, a General, 'a nice old boy and very friendly', who formally surrendered the airfield. To prove it, he gave Nation a receipt, tearing off a page from a pad and writing on it '*La base de Blida est disponible pour l'atterissage des armees allies'* ('Blida base is at the disposal of the allied armies for landing purposes') and signed it, Gaston Menerly.

It was an unusual situation for a Fleet Air Arm pilot to find himself in, but Nation remained chatting with his hosts, somewhat at a loss as to what to do next, until some Commandos and US Rangers arrived and took over the base. Nation then climbed into his Martlet and flew himself and his surrender document back to *Victorious*.

Whilst Hay's Fulmars carried out tactical reconnaissances every few hours, eventually extending their surveys to a radius of 100 miles from Algiers, the Albacores were called up to bomb and silence forts and gun positions overlooking the harbour. At dusk the carriers moved inshore to only four miles off Algiers and were, inevitably attacked. Some fifteen Ju88s were sighted at 22,000 feet but the Seafire CAP could not gain height in time and the bombers dived on *Argus*, who had three Seafires on deck (their pilots strapped in and helpless, spectators of the attack to come). The attackers came down to 500 feet and bombed *Argus* for about four minutes, scoring one hit aft, which damaged the flight-deck port side, blew off one Seafire's tail and damaged others, and scored several other near-misses. Another force of Ju88s attacked Force H, much further out to sea, but were driven off.

By 9 November, Spitfires from Gibraltar had occupied Tafaraoui, and next day Oran surrendered. The carriers' immediate task had been accomplished. Some of them returned to Gibraltar, others remained to continue cover ashore and to escort convoys. It was while escorting a convoy that *Avenger* was torpedoed by U-155 on 15 November. She blew up, with only seventeen survivors. Two days later, a Walrus and RAF Coastal Command attacked and damaged U-331 off the Algerian coast. The U-boat was finally dispatched by a torpedo from an Albacore of 820 from *Formidable*. On the 23rd *Victorious* was on her way to Greenock from Gibraltar when a patrolling Catalina reported that it had damaged a U-boat off the Bay of Biscay. Two Albacores and the destroyer *Opportune* were sent to the spot, where the Albacores' depth-charges sank U-517 and *Opportune* picked up fifty-one of her survivors.

5

1943–4 Critical Years

HUSKY and AVALANCHE

After North Africa, the next objective was Sicily and in July 1943 the two fleet carriers *Indomitable* and *Formidable* were part of the covering Force H for Operation HUSKY, the largest seaborne assault of the war, when the US 7th Army and the British 8th Army landed at several points along the south coast of Sicily. *Indomitable* had thirty Seafires of 807, 880 and 800 Squadrons, with the TBRs of 817 Squadron now augmented to twenty-one Albacores. *Formidable* had a more mixed complement of six Seafires of 885, twenty Martlets of 888 and 889 and 893, and once again, the eighteen Albacores of one of the longest serving (and eventually to be the very last operational Albacore squadron), 820.

The Italian fleet took no offensive part whatsoever, but the Luftwaffe was still active and on the evening of 11 July, the day after the landings, a lone Ju88, which was actually sighted in the moonlight but mistaken for a returning Albacore, flew in and launched its torpedo into *Indomitable*'s port side. (The unfortunate Albacore, when it did return, was given a very hot reception by the guns of Force H.) *Indomitable* had only just returned to service after the repair of her PEDESTAL damage and now she was put out of action for another year.

After Sicily had been secured, *Formidable* made a formal, almost symbolic, entrance into Grand Harbour. It was the first appearance there of any carrier since the departure of *Illustrious* after her long ordeal in January 1941. *Illustrious* herself was present to provide air support for Operation AVALANCHE at Salerno. The first landings on the Italian mainland, at Reggio on the 'toe' of Italy, had been covered by aircraft from Sicily and Malta, and although they were progressing well, it was decided to quicken the pace of the Italian campaign by a further landing in the Bay of Salerno to the north, and seize the major harbour of Naples, in September 1943.

The distance to Salerno meant that only about thirty fighters from Sicily at a time could operate over the beach-head and then only for about twenty minutes on each sortie. The deficiency was made up by the five escort carriers of Force V, commanded by Rear Admiral Vian flying his flag in the cruiser *Euryalus*, who between them could field 107 Seafire IICs: *Attacker*, twenty Seafires of 879 and 886; *Battler*, twenty-one Seafires of 807 and 808; *Hunter*, eighteen Seafires of 834 and 899; *Stalker*, eighteen Seafires of 880; and the repair carrier *Unicorn*, making a rare operational appearance, with thirty Spitfires of 809, 887 and 897. *Illustrious* and *Formidable* were with Force H, *Formidable* with the same aircraft complement as for HUSKY, and *Illustrious* with twenty Martlets of 878 and 890, nine Seafires of 894,

and the first operational Fairey Barracuda squadron, 810, with twelve aircraft.

The Italian surrender was announced while the invasion and covering forces for Operation AVALANCHE were on their way and for a short time there were wild hopes that the operation might be cancelled. These were soon dispelled for good by a night torpedo attack on Force H by some thirty Ju88s. The torpedoes all missed, but they no longer left any doubt that AVALANCHE was necessary.

The Italian battle-fleet sailed south from La Spezia on 9 September to surrender; on the same day, at dawn, Force V off Sicily launched its first twenty Seafires to cover the beach-head. Fresh fighters were launched every hour and the average sortie lasted about eighty-five minutes, so that a continuous patrol of twenty Seafires was maintained. The Seafires flew the phenomenal number of 265 sorties on the first day, by which time the airfield at Montecorvino was expected to be in Allied hands. But by the end of the day, the enemy still held the airfield and the first uneasy suspicions were awakening that the operation was not proceeding as planned.

The aircrews and the ships' companies of the carriers knew only what they heard from the BBC broadcasts, but gradually the news filtered through that the battle on land was not going well. The Seafires were called upon, not for a short, sharp effort as in TORCH, but a prolonged, grinding task which fully exposed the Seafire's limitations as a fleet fighter. The number of sorties flown fell to 232 on the second day, and to 165 on the third. The reason was not enemy action but the steady attrition caused by accidents—pilots new to operational conditions, the very light winds, and the notorious fragility of the Seafire, meant that in five days ten Seafires were lost, thirty written off, and many more were unserviceable because of damaged propellers, hooks, under-carriages, fuselages and engines.

Far out at sea, *Illustrious* and *Formidable* maintained a constant CAP of sixteen fighters, eight over Force H and another eight over the beach-head. They saw no action except for the sole intrusion of a flying-boat which was shot down, prompting one of *Illustrious*'s Fighter Direction Officers, the actor Michael Horden, to declaim aptly Hamlet's speech to the stuck Polonius:

> Thou wretched rash intruding fool, farewell!
> I took thee for thy better.

For the rest, the pilots orbiting endlessly over the fleet could see nothing through a strange meteorological murk of dust and mist which hid the battlefield in a red and purple haze.

The battle was not going well. Montecorvino was finally taken on the third day but had been bombarded so much it was useless and a makeshift airfield was made by flattening fields of growing tomato plants at Paestum nearby. On the fourth day, twenty Seafires from Force V and six from *Illustrious* were flown ashore. They operated from Paestum, amidst perpetually swirling dust-clouds and threatened by shells from both sides until the 15th, when they were relieved by RAF fighters from Sicily. The part they played undoubtedly protected the ground forces from the attentions of the Luftwaffe.

Even so, Salerno was a near thing. A German Panzer division had been resting and re-equipping nearby and the German army launched a furious counter-attack which came close to throwing the landing force back into the sea. Only a bombardment by the heavy guns of the battleship *Warspite* held the perimeter of the beach-head for the Allies at a critical time.

When relieved, twenty-five of the twenty-six Seafires safely took off from Paestum for Bizerta to join their carriers. Such a high survival rate showed that there was nothing wrong with the Seafire as an operational aircraft; it was flight deck conditions which were unsuitable for the Seafire, and always would be.

DRAGOON and the Aegean

In August 1944, after some considerable strategic squabbling amongst the Allies over the merits and demerits of the undertaking, Allied troops landed in the south of France in an operation complementary to the great invasion of Europe the previous June. Operation DRAGOON was one of the least dangerous and most enjoyable assaults of the war and appropriately it took place at the height of the summer holiday season on the French Riviera, at such unwarlike landing beaches as St Tropez, San Raphael and Cannes. These were once again beyond the reach of shore-based air power and fighter cover was provided by the nine escort carriers of Task Force 88, commanded by Rear Admiral Sir Tom Troubridge, flying his flag in the cruiser *Royalist*. TF88 was a joint Anglo-US force, with Troubridge himself commanding Task Group 88–1 of *Attacker* with twenty-four Seafires of 879 Squadron, *Emperor* with twenty-two Hellcats of 800 Squadron, *Khedive* with twenty-six Seafires of 899, *Pursuer* with twenty-four Wildcats of 881, and *Searcher* with twenty-four Wildcats of 882. Task Group 88–2 was commanded by an American admiral, Rear Admiral C.T. Durgin, USN, flying his flag in USS *Tulagi*, with USS *Kasaan Bay*, *Hunter* with twenty-four Seafires of 807 and one Swordfish, and *Stalker* with twenty-four Seafires of 809. Eight Seafire LIIIs each of 807, 809 and 879 had been disembarked that summer to operate ashore at Orvieto as 'D' (Naval) Fighter Wing with the Desert Air Force.

The composition of the covering fleet for DRAGOON showed the immense flexibility of air and sea power then available to the Allies. *Khedive* had just completed her working up; *Hunter*, *Stalker* and *Attacker* had already been operating for some time in the Mediterranean; *Emperor*, *Searcher* and *Pursuer* had just come south from operations off the Norwegian coast. The fleet as a whole comprised British, American, French, Polish and Greek warships and included two of the smallest aircraft carriers ever: two Landing Ships (Tank) with miniature flight decks to operate Piper Cubs for spotting.

Spotting was indeed one of the major tasks for the aircraft. The Germans were fully occupied with the campaigns in Russia and in northern Europe and the 100 or so aircraft of the Luftwaffe remaining in the south of France were not expected to interfere seriously with the landings, nor did they. The fleet fighters' tasks were to provide air cover for the fleet and the beaches as usual, to give close support to the ground forces and carry out reconnaissance flights, for which most of the squadrons had received special training, and to spot for ship bombardment.

The first troops went ashore in the early hours of 15 August, and although it was holiday weather, with light winds and extreme heat above and below decks in the little carriers, only thirty-five of the 200 naval aircraft taking part were written off in making some 2000 landings in the next five days. Enemy air effort was almost non-existent, and the only serious action was some air combat on the 19th over Toulouse, when 110 Seafires and Hellcats swept over the town and destroyed five German aircraft, Junkers Ju88s and Dornier Do217s. Of the 1826 sorties flown, less than a third were CAPs over the ships and the beaches. The rest were made in offensive sweeps inland up the Rhone valley and as far as Avignon, in support of the advancing ground forces. Enemy transport was shot up on the roads, rolling stock was strafed, bridges and railway stations bombed. Some twenty tanks, sixty-four railway trucks and 190 motor vehicles were destroyed or damaged, eighteen roads and fourteen railways were cut, and seven small ships, including one Motor Torpedo Boat, were sunk. Against those, twenty-three aircraft were lost, and nine pilots.

The statistics of DRAGOON—nine carriers, over 200 aircraft, and nearly 2000 sorties—illustrate the astounding growth of the Fleet Air Arm. The problems of operating too few aircraft, from one single beleaguered carrier, were replaced by the new problems of operating several carriers and scores of aircraft at a time. Taking off and landing on in quick time by squadrons, marshalling and forming up in the air, radar interpretation and fighter direction of several aircraft groups at different heights and courses, were all taken to new lengths, embellishing and improving upon old skills.

In September the seven carriers from DRAGOON moved into the Aegean for final 'mopping up' operations. There was very little enemy aircraft activity and most of the fighters' effort was devoted to strafing and bombing coastal defences, troop movements, and coastal vessels such as schooners, caiques and lighters, which the German forces in the Aegean were using to evacuate the islands. Although the work of convoy and cover, submarine search, reconnaissance, air-sea rescue, ferry and ground liaison, went on until the end of the war, these Aegean operations were the last major undertakings by the Fleet Air Arm in the Mediterranean.

Victorious in the Pacific

By October 1942 the US Navy's carrier strength had sunk so low during the campaign on Guadalcanal that the US Chiefs of Staff asked the Admiralty for the loan of one, and better still two, aircraft carriers to operate in the Pacific. The request was awkwardly received in London and for a time there was a definite breakdown in communications between the Allies. The Admiralty did not at first realise the Americans' predicament but in any case had many other preoccupations world-wide, including the constant battle in the Atlantic, the restart of the Arctic convoys, and the outcome of TORCH which was still in doubt. It was not until December that one carrier, *Victorious*, could be spared. She reached Pearl Harbour in March 1943 but her aircrews and ship's company had to be trained in American practice and procedure, and she did not join the fleet until May, when she had three fighter squadrons, 882, 896, and 898, with thirty-six Martlets, and fifteen Avengers of 832 Squadron embarked.

With the US carrier *Saratoga*, *Victorious* was part of Task Group 36–3 covering the landings in New Georgia in June and July 1943. Relations between the two ships were very cordial. Captain Mackintosh went to sea in *Saratoga*, Rear Admiral Ramsay, USN, commanding carrier forces in the south-west Pacific, and Captain Mulliner, USN, of *Saratoga*, visited *Victorious*. Aircraft 'cross-operated' from each others' flight decks, six Avengers and twelve Martlets from *Victorious* once being exchanged for eight Avengers, six Douglas Dauntlesses and twelve Wildcats from *Saratoga*. The Avengers of 832 embarked in *Saratoga* for four weeks in July, one of the rare occasions when British aircraft operated from an American carrier. *Victorious* herself returned home via the west coast of the United States in September.

The Battle of the Atlantic: Fighter Catapult Ships, CAM and MAC Ships

The loss of *Ark Royal*, one of the last and easily the most famous link with the pre-1939 Fleet Air Arm, was like the end of an era. But a new era had already begun that summer of 1941, with the introduction of the first converted Fighter Catapult Ship. It had at last been realised that aircraft were the most flexible and efficient weapons against U-boats, either by attacking directly with explosives, or by guiding escort vessels with explosives to the spot, or at least by neutralising the U-boat by forcing it down, and keeping it down, until the convoy had passed.

The convoys were the key. It was almost useless, and could sometimes be disastrous for carriers to plunge about the oceans searching for U-boats. That tactic had cost the Navy *Courageous*. U-boats were far better engaged near their prey, that is, in and around the convoys. Continuous air cover had to be provided above and in front of the convoy as it went along, either by long-range shore-based aircraft, or by aircraft from a carrier inside and part of the convoy itself. It was unnecessary and wasteful to use a fleet carrier for this. What was needed was some smaller and more dispensable kind of carrier, something cheap and cheerful, as the saying went. This was the role that the escort carriers, the 'Woolworth carriers' fulfilled so well.

But, at first, there were no escort carriers and the Admiralty had to improvise. The old seaplane carrier *Pegasus*, with three Fulmars of 807 embarked, had covered convoys from December 1940. In the spring of 1941, four ex-merchant ships, *Patia*, *Springbank*, *Ariguani* and *Maplin*, were converted for service as Fighter Catapult Ships, each being fitted with a catapult ramp and one Fulmar or Sea Hurricane.

Patia was sunk in April 1941, before embarking her aircraft, but the others, each with a Fulmar of 804 Squadron, sailed with convoys from May 1941 onwards. Their chief opponents were not the U-boats so much as their partners, the long-range four-engined Focke-Wulf Kondor Fw200 bombers. After the fall of France, the Luftwaffe was able to use airfields on the west coast which gave the searching Kondors a range of more than 700 miles out into the Atlantic, where they could sight convoys which were out of range of Allied aircraft and which had no surface escorts with them, and so report their positions to prowling U-boats.

In time, some fifty merchant ships, each carrying one Hurricane, though still flying the Red Ensign, were converted and commissioned as Catapult Armed Merchant Ships, or CAM ships. The RAF Hurricane pilots (all but one, who was a naval pilot, in *Michael E*) were a special breed. Flying from a CAM ship needed a peculiarly cold-blooded kind of courage. When a shadowing Kondor was sighted the Hurricane was launched, engaged the enemy above or near the convoy and then, unless there was a (very rare) shore airfield to divert to, the pilot had to bale out or ditch and trust to be picked up out of the water.

Springbank was sunk while escorting a Gibraltar convoy on 27 September 1941, but meanwhile the first success by a fighter catapult ship had been scored by *Maplin*, whose Sea Hurricane shot down a Fw200C on 3 August 1941. The pilot was Lt R.W.H. Everett, RNVR (who had also piloted Gregalach to win the Grand National in 1929). He got within one and a half miles of the Focke-Wulf before it noticed him. He intercepted his quarry after nine minutes' flying and ranged up alongside at 600 yards and slightly above it. 'When my machine was slightly ahead of its starboard quarter the stern gunner opened fire,' he wrote in *Fleet Air Arm*. 'These rounds passed underneath or fell short of the Hurricane. It took quite an appreciable time to get abeam and the for'ard cannon was also firing—again the rounds passed underneath or short. The Focke-Wulf then turned sharply to port, but seemed to change its mind and turned back on its original course. By this time I had reached its starboard bow and their machine-guns opened up as well as the for'ard cannon. I did a quick turn to port and opened fire just abaft the beam. I fired five-second bursts all the way until I was forty yards astern of the enemy. Another short burst at this range and my guns were empty. I noticed pieces flying off the starboard side of the Focke-Wulf and it appeared to be alight inside the fuselage. I broke away to port at thirty yards. My windshield and hood were covered with oil and I quickly jumped to the conclusion that my engine oil system had been hit.'

In fact the oil came from the Focke-Wulf which dived into the sea a few seconds later. Everett's forward view was still obscured by oil and 'my one idea was to get down while I still had charge of the situation. I made two rather half-hearted attempts to bale out, but the machine nosed down and caught me half out. I changed my mind and decided to land in the sea near HMS *Wanderer*, and did so. The ship sent a boat and I was extremely well looked after.' (Everett was killed in January 1942 while ferrying an aircraft from Liverpool to Belfast.)

Maplin was the last Fighter Catapult Ship in service, being returned to normal trading on 30 June 1942. The CAM ships, supplied with Sea Hurricanes and handling parties by the RAF Merchant Ship Fighter Unit, survived much longer. Between May 1941 and August 1943, CAM ships made 170 round voyages with convoys across the Atlantic, to Gibraltar and to the Arctic. Twelve CAM ships were lost.

Only eight operational launches were made, by seven CAM ships (*Empire Morn* launching twice) but achieved a very high striking rate, destroying six enemy aircraft and damaging another three. The last two, *Empire Tide* and *Empire Darwin*, sailed in convoy from Gibraltar on 23 July 1943, and on the 28th, Hurricanes from these two ships signed off the wartime record of the CAM ships by shooting down two Kondors.

A further development was the Merchant Armed Carrier, or MAC ship. This was a tanker or grain-ship with a rudimentary flight-deck built over and around its normal superstructure. The 'grainer' had a lift and a small hangar and could operate up to four Swordfish. The tanker normally had a longer and narrower flight-deck, but had no hangar and could operate a flight of three Swordfish. Both types had a safety barrier and two arrester wires, and could carry ninety per cent of their normal cargoes with aircraft embarked. The MAC ship Swordfish were provided by 836, and 840 Squadrons, and by 860 Squadron of the Royal Netherlands Navy, all based at Maydown in Northern Ireland.

By normal operating standards, the MAC ship flight-decks were minute, being not much wider than the Swordfish's wing span and so short that a Swordfish loaded with depth-charges or rockets required RATOG (rocket assisted take-off gear). The first MAC ship was *Empire MacAlpine*, commissioned in April 1943. Nineteen MAC ships in all, six 'grainers' and the rest tankers, saw convoy service from the summer of 1943 onwards. Some convoys in 1944 had four MAC ships in company. MAC ships' Swordfish flew over 4000 sorties in all and made twelve attacks on U-boats. They sank no U-boats, but no convoy with a MAC ship ever lost a ship to U-boats.

Successful though they were (especially at reassuring the other ships in convoy) the MAC ships themselves were really only makeshifts, introduced because the real solution, the auxiliary or escort carrier, was slow in coming into service. These were small ships, converted from merchant ship hulls, but still genuine aircraft carriers, although, strictly speaking, the first auxiliary aircraft carrier was yet another improvisation, a kind of MAC ship on her own. However, she and her successors did have a radical effect on the progress of the war at sea.

The Escort Carriers

Her name was *Hannover*, an ex-German liner captured as a prize by the light cruiser *Dunedin* off San Domingo in February 1940. She was taken in hand for conversion into an auxiliary carrier, her masts, funnel, upper bridge and other superstructure being removed and replaced by a single flush deck, 368-feet long and sixty wide. First named *Empire Audacity*, she was commissioned as HMS *Audacity* early in 1941. She had no hangar, and all work and maintenance on her six Martlet fighters had to be done in the open. She had two arrester wires, a safety barrier and, just aft of the barrier, a third wire, with no hydraulic pull-out, which the pilots called the 'Jesus Christ' wire.

Her squadron, 802, flew on from Campbeltown in September 1941 and took the ship to their hearts—not least because of her superb accommodation. The Captain, Cdr D.W. McKendrick, an old Swordfish pilot himself, insisted that the ship's staterooms, which remained from her days as a liner, be allocated to aircrew. So the pilots of 802 found themselves with single beds, bathrooms attached, in a sybaritic luxury unequalled anywhere else in the fleet.

At sea, *Audacity* revolutionised the theory and practice of convoy defence, although 802's Commanding Officer, Lt-Cdr J.M. Wintour, RN, was lost in particularly distressing circumstances on 8 November in defence of convoy OG 76. He had engaged a Fw200C astern of the convoy and, having set it on fire, ranged up alongside it, believing its guns to be silenced. But one of its guns opened fire and hit the Martlet's belly beneath the cockpit, as Wintour banked to get away. The Martlet's R/T had been switched to the ship's broadcast and Wintour's dying cry was relayed right through the ship. As some consolation, his wingman, Sub-Lt D.A. Hutchinson pursued and destroyed the Kondor.

Audacity proved her worth beyond all question during the passage of the homeward bound convoy HG 76, which sailed from Gibraltar in December 1941. The convoy's surface escort was the 36th Escort Group with destroyer reinforcements, led by that formidable Commander (later Captain) F.J. Walker, who was already becoming the scourge of the U-boats.

Once beyond the range of aircraft from Gibraltar, the ships of HG 76 had to rely on their own efforts until they came within range of home-based Coastal Command again, but the combination of Walker's well-drilled escorts and 802's Martlets, delivered a devastating shock to Donitz's U-boats and to the Luftwaffe.

Audacity sailed from Gibraltar, on 14 December 1941, with only four Martlets serviceable, after her exertions in defence of OG 76. On the 17th a Martlet on dawn patrol reported a U-boat on the surface twenty miles on the convoy's port beam, and attacked. The U-boat's fierce return fire shot down the Martlet and the pilot, Sub-Lt G.R.P. Fletcher, RNVR (who had shot down a Fw200C on *Audacity*'s first trip in October) was killed. But the escorts came up and sank the U-boat, U–131.

So began a four-day air and sea battle. The convoy was detected by Kondors and as many as nine U-boats closed for the attack. Two merchant ships and the destroyer *Stanley* were sunk, but two Fw200s were shot down and three U-boats were sunk around the convoy. But on 21 December, during a hectic anti-submarine battle, *Audacity* herself was torpedoed. She had steamed to starboard of the convoy, against Walker's advice. With her diesel engines *Audacity* could make only 14 knots, and Walker preferred her to stay within the convoy screen, or at least steam to port, the safer side. Walker afterwards reproached himself; McKendrick was senior to him, but Walker was the escort commander and he felt he should have insisted. Some of *Audacity*'s wardroom expected to be torpedoed that night; as Lt Donald Gibson, who had flown out to replace Wintour, later said, 'it came with the coffee'. *Audacity* was brilliantly illuminated by a flare after a ship had been sunk in the rear of the convoy. Plainly silhouetted, *Audacity* was hit by one torpedo and slowed down, beginning to settle by the stern. Her rudder jammed, and she came to a stop whilst a weird sight, reminiscent of the Flying Dutchman, appeared close by: a U-boat glowing brightly with St Elmo's Fire, passed only a cable away to port, and fired two more torpedoes which were too much for *Audacity*. The fore part of the ship was blown off and the rest sank ten minutes later. McKendrick was not amongst the survivors, being washed away while he was trying to get into a rescue whaler.

It was a hard war for 802 Squadron. They had three carriers sunk under them. They had been in *Glorious* and were also in *Avenger* which was sunk on 15 November 1942. Of five wartime Commanding Officers of 802, Gibson was the only one to survive. The other four were all killed in action whilst leading the Squadron.

The first real escort carrier, though again a conversion, was HMS *Archer*, commissioned at Norfolk, Virginia on 17 November 1941, with Captain J.I. 'Streamline' Robertson, who had been *Illustrious*'s Commander (Air) in the desperate Mediterranean days, as her commanding officer. The Americans designated escort carriers AVGS, and *Archer* (ex *Mormacland*) was numbered BAVG1, (B for British). Five more C-3 merchant hulls were converted, becoming *Avenger* (ex *Rio Hudson*), *Biter* (ex *Rio Parana*) and *Dasher* (ex *Rio de Janeiro*). The fifth, USS *Charger*, was retained in the United States for deck-landing training. They were of some 9000 tons displacement, could make about 17 knots, and carry fifteen aircraft. They had a hangar, one lift, and a wooden flight deck which was 410-feet long by eighty-seven-feet wide.

Archer was evidently laid down during an evil star, because almost everything that could go wrong with her did so. On her shakedown cruise she had trouble with her catapult, her gyro compass, her main engines, her steering gear and her radar and, on 12 January 1942, she collided with a Peruvian steamer and struggled back to harbour, with a jury rudder, under tow.

Ready again in mid-March 1942, *Archer* sailed for Sierra Leone and another chapter of accidents. Her main engines gave constant trouble, her ship's boats failed to start, her fresh-water distilling plant produced either rusty or oily water, and her high-pressure air compressors would not work. *Archer*'s deck officers and most of the seamen and technical ratings were Navy men but her engineer officers and her stewards were Merchant Navy, engaged under the special T124X articles. They solved a serious manpower problem for the Royal Navy, but they were restive under naval discipline and in any case were permitted to lead a Merchant Navy life, even though serving in a warship.

To problems below were added accidents above. There were several serious deck-landing crashes and some fatalities. A Swordfish dispatched with a message to Ascension Island was fired on by the islanders and the pilot was wounded. A 250-lb bomb exploded in the walkway beside the flight deck, killing nine men and injuring ten more, blowing a hole in the hangar and setting fire to one aircraft. By July 1942, *Archer* had limped back to America, having broken down in mid-Atlantic, with twenty-six U-boats reported in her vicinity and all the aircrew down with 'flu. Unsurprisingly, there were a number of bitter little ditties about *Archer*. One of them was quoted in Kenneth Poolman's *Escort Carrier 1941–45*:

> There'll always be an *Archer*, with
> aircraft always there—
> Providing all the engines work, she'll
> get them in the air.
> There'll always be an *Archer*, as long
> as there's a berth,
> A lease-lend gift from Heaven, but no
> bloody good on earth.

However, *Archer* had her hour, one fierce hour and sweet. It came in May 1943, when repaired and once again operational, *Archer* and the 4th Escort Group were on convoy escort duty in the Atlantic. *Archer* and her accompanying destroyers and sloops operated as a separate group, transferring from one convoy to another, to provide air and sea escort, as the situation developed. May 1943 was a crucial time in the battle of the Atlantic, when the U-boats received a serious defeat, their first of the war.

Sunday, 23 May was a day of intense activity, with several U-boats reported, either by sighting, by sonar detection, or by radio direction finding. *Archer*, with the destroyers *Faulknor*, *Onslaught* and *Impulsive* and the sloop *Pelican* had joined the home-bound convoy HX 239 two days earlier, having already been in action with two other convoys. *Archer* had nine Swordfish of 819 Squadron and three Martlets of 892 embarked; the two types of aircraft worked together, the Swordfish attacking with depth-charges or rockets, with the Martlet's gunfire in support.

Shortly after 9 a.m., one of *Archer*'s Swordfish sighted the 'feather' of a U-boat periscope in the water and dropped depth-charges which badly damaged the submarine though they did not sink her. Another Swordfish and a Martlet, coming to assist in this attack, sighted a second U-boat, a big dirty-white 'milch cow', on the surface and drove her down with depth-charges and machine-gun fire.

Finally, an hour later, a Swordfish piloted by Sub-Lt H. Horrocks, RNVR, with Sub-Lt W.W.N. Balkwill, RNVR as observer and Leading Airman J.W. Wick as TAG, sighted a U-boat on the surface making about 15 knots. Horrocks flew through cloud to close his target and his four salvoes of rockets caught the submarine completely by surprise. The U-boat, unable to dive because of hits on her hull, heeled over, and began gushing oil. A Martlet, summoned to help, arrived within a minute and fired 600 rounds into the conning tower, killing the U-boat captain. At 10.50 the U-boat crew came on deck to abandon ship, as the submarine, U-752, sank beneath them.

Archer later went to the Gareloch and acted as a stores and accommodation ship there and in Belfast until she finally paid off in March 1945.

Biter also had great success in the Atlantic against U-boats, her 811 Squadron Swordfish sharing U-203 with *Pathfinder* on 25 April 1943, and U-89 with *Broadway* and *Lagan* on 12 May. But the Admiralty had reservations about American standards of ship stability and fire precautions. The Admiralty's caution was justified when *Avenger*, after one torpedo hit, sank after a subsequent internal petrol explosion, with only seventeen survivors. On 27 March 1943, in the Clyde, another of the class, *Dasher*, blew up and sank with the loss of 378 lives, almost certainly due to a spark or a cigarette igniting petrol fumes. Escort carriers spent many months in British dockyards, having major alterations carried out, before they were admitted into service.

In December 1941 a great escort carrier building-programme was put in hand. Twenty C-3 merchant hulls were allocated for conversion, ten of them becoming the US *Bogue* Class, and ten becoming the Royal Navy *Attacker* Class; they were of 14,000 tons, had a speed of 17 knots, a flight deck 450-feet long and 100 wide, two lifts and could carry sixteen fighters and twelve torpedo-bombers. In April 1942, twenty more C-3 hulls were allocated and all went to the Royal Navy, becoming the *Ruler* Class. Four escort carriers, all converted from merchantmen, were built in the United Kingdom: *Activity*, commissioned in September 1942, and *Vindex*, *Nairana* and *Campania*, all laid down and completed in 1943.

The small 'Lend-Lease' escort carriers were nicknamed 'Woolworth Carriers' and the name stuck although, according to the magazine *The Aeroplane* which sent a man to look round *Battler* when she first arrived in the United Kingdom from the States, this was 'a gross misnomer and tends to give a dog a bad name before it has started to bark. These ships are not cheap imitations.' In *Flat Top*, F.D. Ommanney describes their somewhat odd appearance: 'Try to imagine a ship cut off along the line of her deck and covered by a rectangular box with a flat overlapping lid. You climbed up the gangway and entered the box by a hole in the side. There were no funnels or masts such as a ship ordinarily has and the smoke came belching untidily out from two tubular vents, one each side under the lid. On the starboard side a little irregular house was perched. It bristled with an assortment of masts and gadgets and was known as "the island". It was all there was in the way of upper-works. It contained the bridge, chart-room, wheel-house and all that impressive nerve-centre of a ship that is usually forbidden to strangers and labelled: "No admittance. Ship's Officers and Crew Only".

'The lid which covered the whole ship from stem to stern was the flight-deck, a small ship-borne landing ground. Under it, inside the rectangular box which seemed to have been clamped upside down over the ship, was a great echoing space, the hangar deck, where the aeroplanes were stored and serviced. It was really a floating garage. It communicated with the flight-deck above by means of two lifts, one forward and one aft, each large enough to carry one aeroplane. And from the hangar you went down through narrow hatchways, as though into a luminous metallic underworld into the living and working quarters in the hull of the ship underneath.'

In that ship underneath, the ship's company found American standards of comfort which were eye-openers to sailors used to RN ships. The men no longer ate in their messes, each with a cook of the mess to serve up the meal, but in a central mess-hall with a cafeteria system, and not off the old china mess traps but shiny metal trays with food compartments. There were bunks instead of hammocks and a barber's shop and a properly equipped laundry (which, with misplaced Spartan zeal, was not used in some ships, the sailors being made to do their own 'dhobeying' in the traditional way). The officers had single or two-berth cabins, each with cupboards, desks, wardrobes and even a private safe for belongings. The hangars were light and airy and were fitted with facilities for film-projection.

The British escort carriers were of sterner stuff, rivetted instead of welded, with an 'enclosed' hangar which was smaller and seemed much darker than in an American ship, one lift instead of two, and a steel flight deck which was longer, at about 490 feet, but narrower, at seventy-seven feet. The wider American deck made deck handling and stowage of aircraft above much easier. The British-built carriers had a slight speed increase of about two knots.

In almost all the escort carriers, the engine-room, cooks' and stewards' departments were manned by Merchant Service men on T124X articles. This made difficulties in maintaining discipline, for an offender could often point to an exonerating clause in his Board of Trade agreement. Many of the engineer officers, Sub-Lieutenants (E) RNR, repudiated officer status, insisted on being known as 'Junior Engineers' as though they were in a merchantman, and sometimes operated a kind of snobbery in reverse by appearing in the wardroom in greasy overalls.

By 1944 there were few Atlantic convoys not accompanied by an escort carrier, equipped with a 'mixed' squadron of Swordfish and fighters, normally Wildcats but occasionally Seafires. The aircraft flew hundreds of hours over thousands of square miles of ocean. They very rarely saw an enemy, but they did have their successes, each one like a brilliant ray of light in a grey and seemingly interminable expanse of duty, as featureless as the sea itself. Swordfish of *Fencer*'s 842 Squadron sank U-666 in the Atlantic on 10 February 1944, and Swordfish of 825, Esmonde's old Squadron, flying from *Vindex*, shared U-653 with *Starling* and *Wild Goose* on 15 March and U-765 with *Aylmer*, *Bickerton* and *Bligh* on 6 May 1944.

A very typical attack, actually an amalgam from the facts of several similar attacks, was entertainingly and authentically written up, as though from a Swordfish squadron linebook by John Moore in his *Escort Carrier*: '25. Today, being Sunday, the better the day the better the deed, the ship got her first kill. F. in Miss Blandish on routine patrol just after dawn spotted a surfaced U-boat which apparently took some time to recover from its astonishment at meeting a single-engined biplane in mid-Atlantic. F. made his sighting report and attacked; at least we supposed he had attacked, for we heard some machine-gun fire on the R/T, then silence. F. did not answer any calls after that, we heard no more of him, and he is "presumed shot down".

'Two Swordfish and one Seafire flew off at once (Squadron Commander and Seafire Flight-Commander pinching this trip, of course, as they always do when there's anything worth having!). They found the submarine still on the surface, going round in aimless circles, having no doubt been damaged by F. before it shot him down. Squadron Commander said on the R/T "Softly, softly, catchee monkey" and they cruised round for a minute or two just out of range. Then the Seafire went down and let him have it with m-g's and cannon. The Swordfish saw strikes on the conning tower, and the gun crews tumbling off the deck. Then they attacked with depth-charges. The explosion seemed to heave the submarine's bow right out of the water. After that it disappeared altogether for a few moments, then suddenly came up again stern-first. The stern stuck up out of the white boiling sea like a sharp black rock. A beautiful sight. It sank very slowly, leaving a few survivors swimming about. These were later picked up by one of the sloops. They said the Seafire's first burst had killed their Captain. Every time a coconut.'

The best comment on the escort carriers' work in the Atlantic was the tribute (quoted by Kenneth Poolman) of the Commodore of Convoy HX 258 in October 1943. 'I have never had such a splendid escort, surface or air, before. We had in the chief danger area fourteen surface escorts, an aircraft carrier (actually *Tracker*) and three Liberator shore-based aircraft over us all the daylight way across. My convoy was tickled to death and felt like a dog with two tails; one could sense the rise in morale. It was no doubt due to this great strength in escort that we were unmolested.'

The Squadron Linebook

'Line' is very probably short for 'lineshoot' wrote John Moore—a term very probably borrowed by the Fleet Air Arm from the RAF. The original 'line' goes something like this: "There was I . . . on fire . . . upside down . . . in cloud . . . nothing on the clock except the maker's name . . . and *still* climbing!' Somebody at some time had the idea of collecting the best and least probable 'lines' into a book which then became known as the linebook. Almost every squadron during the war, and since, has kept a linebook although many have not survived. There was no standard format; they were generally large, blank-paged, of a scrap-book type, in which were pasted photographs, cartoons, apposite newspaper headlines and cuttings, invitation cards, theatre programmes, ship's daily orders, signals—anything that seemed appropriate to the squadron member who was deputed to keep the linebook.

To give a solemn description of an unsolemn phenomenon, the squadron linebooks were pieces of contemporary history written by those who were making it. Even now, years later, surviving wartime linebooks have an almost eerie feel of authenticity, because they give 'accounts of the squadron's battles and the squadron's parties; there were various versions of the squadron's unprintable songs; there were photographs of the squadron's more spectacular prangs. It was a curious hotch-potch in which blitzes and popsies, bombs and beer, indeed, life and death, were treated with an equal lack of solemnity.'

As items of naval history, linebooks are now priceless. As John Moore wrote of one Swordfish squadron's linebook it 'was battered, torn and in places almost illegible. It had accompanied the squadron across half the world, and it recorded impartially and at times ironically their triumphs and disasters during nearly three years of war. It was stained by sea water; it was scorched by fire, having been rescued with difficulty when a dispersal hut was hit by an incendiary bomb at Malta. It had known the sand and the mud of the Western Desert, it had been evacuated from Crete, and saved from the torpedoed *Ark Royal*. Its value, in the eyes of the squadron, was greater than that of the *Codex Sinaiticus*.

'It contained, in the first place, a list of every pilot, observer and air gunner who had served in the squadron; in many cases with their photograph as well. There were more than 200 names. Opposite some of them were the initials "P.O.W." Opposite many, many more was "R.I.P."

'The rest of the book was filled with the Squadron Diary, dating from its formation in 1940, which told a story of Mediterranean war, of Crete, Malta and Libya, that has never been told on any printed page, nor can be told until the war is over.'

The Arctic

Whatever the political and moral justifications for the convoys which from 1941 onwards took supplies to Russia round the North Cape, as naval undertakings they were always tactically unsound. Until very late in the war, the Arctic convoys involved risks which could be, and should be, taken once and possibly even twice, but not repeatedly. The merchant ships and their escorts had to make their way, at the speed of the slowest ship, to and from Russian ports along a route which was often restricted by polar ice and always within easy range of enemy aircraft, U-boats and surface ships, but where Allied heavy ships could not protect them. To the violence of the enemy were added the dangers of the appalling Arctic weather and the hostility of a suspicious and ungrateful ally.

The first convoy, escorted by the aged carrier *Argus*, sailed as early as August 1941, but it was not until March 1942 that Albacores from *Victorious*, at sea with the Home Fleet to provide deep cover for the Russian convoy PQ12, carried out a strike which was unsuccessful but had a profound and far-reaching effect upon German naval tactics in the Arctic.

On 9 March, the German battleship *Tirpitz* had been reported at sea and in fact was steaming south towards Narvik, having missed the convoy. At 6.40 a.m. that morning *Victorious* flew off a search of six Albacores and followed them at 7.50 with a striking force of twelve Albacores, of 817 and 832 Squadrons, led by Lt-Cdr W.J. Lucas. As they went, the Commander-in-Chief, Admiral Sir John Tovery, signalled: 'A wonderful chance which may achieve most valuable results. God be with you.'

At 8 a.m. one of the searching Albacores, piloted by Sub-Lt W. H.C. Brown, RNVR, sighted a suspicious vessel which he identified as *Tirpitz* in a position about eighty miles from the fleet. Brown was joined by another Albacore, whose observer Sub-Lt (A) T.T. Miller confirmed two minutes later that the vessel was *Tirpitz*, with one destroyer (*Friedrick Ihn*), steaming south at high speed. *Tirpitz* herself sighted the two shadowers about 8.15, launched her own Arado seaplane and turned east at 29 knots.

Lucas, slowly overhauling his target, actually sighted *Tirpitz*, twenty miles away to the south-east, at about 8.40, and the strike altered course to close, climbing at the same time to 3500 feet in cloud. The Albacores had a good radar contact of *Tirpitz* and were firmly in touch from about sixteen miles' range. Lucas intended to stay in cloud, tracking by radar, until his force had arrived in front of *Tirpitz* and up wind, and then attack.

But the wind at 3500 feet was easterly, 35 knots, dead in the attackers' teeth, and the Albacores were only catching up their target at a relative speed of about 30 knots. At 8.50 Lucas ordered the four sub-flights to act independently, possibly because they were running the risk of icing up at that altitude, on that day. But from here the attack went wrong. Lucas sighted *Tirpitz* through the cloud and, because she did not open fire, assumed he had taken her by surprise and so, at 9.20 gave the order to attack. In fact, Lucas sacrificed his precious height only to find himself almost on top of *Tirpitz* and then astern of her, and having to labour after her, dead into the wind. The other sub-flights also faced a sternchase upwind against a thoroughly alerted target, who soon turned north and then east again to evade the torpedoes, whilst engaging the Albacores with a very heavy and accurate flak barrage which shot two of them down.

None of the torpedoes hit, and evidence from cameras in the Albacores showed that they had been dropped outside effective ranges. However, *Tirpitz*'s log mentioned the great 'determination and dash' of the attackers, dropping their torpedoes at ranges of 400 to 1200 yards, one of the torpedoes missing astern by only ten yards. Some of the Albacores flew in over *Tirpitz*'s intense AA barrage (she mounted eighty-eight guns of 20-mm and 150-mm calibres) to machine-gun her superstructure. Lt Karl Theodor Raeder, *Tirpitz*'s Gunnery Officer, saw one of the crashed Albacores in the water. 'I remember seeing one of the airmen sitting helpless on the top wing. In that ice cold water there could be no doubt that in a few minutes he would be dead. But we had just survived one attack, another might be in the offing, and there could be no question of stopping the battleship to pick up an enemy airman.'

Untouched, *Tirpitz* anchored in Narvik the next day, while the Home Fleet, bitterly disappointed, returned to Scapa. The Admiralty, the fleet, *Victorious* and the Albacore squadrons all knew that a 'wonderful chance' had been lost (which, in fact, never came again in the whole war). Captain Bovell wrote: 'No one is more disappointed than the crews of the aircraft who took part in the attack; it was the chance they had dreamed of and prayed for.'

In mitigation, one should remember that Lucas had taken over his squadron only a little time earlier and had never led them in the air before. Neither had he had much recent experience of torpedo attacks, which needed constant and frequent practice. He and his crews had been put to 'a very severe, even unfair test'. Perhaps the fleet and *Victorious* were misled by the success of the torpedo attacks against *Bismarck*, where it is quite possible that the skill of the aircrews, and sheer luck, compensated for lack of training. However, Lucas' attack had an effect on the enemy as powerful as though he had actually hit and disabled *Tirpitz*. Hitler forbade her to put to sea again when there was the slightest chance of an aircraft carrier being pitted against her, and ordered the completion of work on the *Kriegsmarine*'s sole carrier *Graf Zeppelin* (which was, in fact, never carried out).

However, the Admiralty was unaware of this restriction and when *Tirpitz* put to sea again in July 1942 to attack the convoy PQ17, the mere whisper of her possible presence caused a disaster for the Allies. *Tirpitz*'s sortie was brief and she never came near PQ17, but from London the convoy was ordered to scatter. Deprived of mutual support and covering gunfire from their escorts, the unfortunate merchantmen were massacred by U-boats and the Luftwaffe. Twenty-two of the ships in PQ17 were lost. Churchill called it one of the most melancholy naval episodes of the whole war, and Stalin hurried to rub salt in the wound by asking if the British Navy had any sense of glory.

The next Arctic convoy, PQ18, did not sail until September 1942, because of the pressing need to supply ships for TORCH and to wait for longer hours of darkness. The convoy of forty ships was strongly escorted by the anti-aircraft cruiser *Scylla*, flying the flag of Rear Admiral Burnett, sixteen fleet destroyers, a CAM ship *Empire Morn*, some trawlers and mine-sweepers and, for the first time in a Russian convoy, an escort carrier, *Avenger*, with twelve Sea Hurricanes of 802 and 883 Squadrons, and a flight of three Swordfish of 825 embarked.

Avenger joined the convoy off Iceland on 9 September, the sight of her greatly cheering the ships in convoy (although her twelve fighters would have to oppose some ninety-two torpedo-bombers and 133 long-range and dive-bombers of Luftflotte V, based in northern Norway). Radio silence and thick fog protected PQ18 until the 13th when *Avenger*'s fighters flew off to drive away enemy shadowers. But while the fighters were away, some forty torpedo-bombers coming in, in Admiral Burnett's words like 'a huge flight of nightmare locusts', sank eight ships in the starboard columns of the convoy. Two more groups of torpedo-bombers made attacks without scoring, and at the end of the day five enemy aircraft had been shot down by the ships' AA fire and one snooping Blohm & Voss Br138 damaged by a Hurricane.

Bitterly, Captain Colthurst in *Avenger* realised he had made a mistake. 'I had not appreciated the hopelessness of sending even four Sea Hurricanes to attack the heavily armed shadowers.' Ironically, *Avenger*'s Hurricanes were early types, IBS, while the merchantmen they were protecting had in crates in their holds much later and faster types, Marks X or XI—an anomaly over which Admiral Tovey protested to the Admiralty.

The next day began splendidly for *Avenger* when, during the forenoon, one of 825's Swordfish and the destroyer *Onslow* made a co-ordinated attack which sank U–589, the first escort carrier U-boat kill in the Arctic. The fighters, too, had a strenuous day. Shadowers hovered around the convoy all forenoon but Colthurst, growing cannier with experience, held his Hurricanes back for the large attacking formations which he knew must come.

They came just after midday, about twenty Ju88s and Heinkel IIIs, and *Avenger* got nine Hurricanes in the air to meet them. 'It was a fine sight,' wrote Admiral Burnett in his report, 'to see *Avenger* peeling off Hurricanes whilst streaking across the front of the convoy from starboard to port inside the screen with the destroyer escort blazing away with any gun that would bear, and then being chased by torpedo-bombers as she steamed down on an opposite course to the convoy to take cover. . . . Altogether a most gratifying action.'

It was indeed. The Hurricanes forced off the first attackers and then turned on a second wave of Ju88s approaching from astern. The air action began to take on the picture of a general mêlée. 'You could see them coming in layers like a wedding cake,' wrote one pilot in *Fleet Air Arm*, 'and as we took off, it looked as though we had about three aircraft to every layer of Huns. Our squadrons had to split up to tackle various bunches of Huns, and eventually I found myself with my section mate—a Petty Officer who was a wizard pilot and a grand fighter—tackling fourteen Junker 88s flying in diamond formation, a pretty hard nut to crack, for if they can keep formation their cross-fire keeps every plane covered.

'However, I made a quarter attack on the leading plane, then swung away straight at one of the planes on the side of the diamond. At the last second I flicked underneath him; he got the wind up and pulled the nose of his plane hard up and the Petty Officer, flying just on my starboard wing, gave him a lovely burst which put paid to his account. The formation broke up, and there was a lovely scrap all over the sky.

'That sort of thing went on all day. As soon as we were out of ammunition or petrol, we dived down to the carrier, landed, rearmed and refuelled and took off again. My lunch was a gulp of cold tea. Our squadron made seventeen sorties that day.

'I saw two Heinkels, having launched their torpedoes, flying along the side of the carrier dead level with the bridge. The gunners waited till they were only a few yards away, opened up—and the sky and sea were full of bits of Heinkel, an amazing sight.' Three Hurricanes followed their enemy through the fleet barrage and were shot down, but all three pilots were rescued.

The enemy made a dead set at *Avenger* herself, who survived several bomb near-misses and evaded some seventeen torpedoes. But, at the end of the day, she was still there and her Hurricanes claimed five certain, three probables and fourteen damaged. As the pilots said, 'our job was to make the Huns miss, not to become aces'. On the 17th, *Empire Morn*'s Hurricane was launched and shot down a Heinkel torpedo-bomber. The pilot, Flying Officer A.H. Barr, then drove off other attackers with dummy attacks and finally landed at Keg Ostrov, near Archangel, 240 miles away with four gallons of fuel remaining. PQ18 lost a total of thirteen ships (three to U-boats), and four Hurricanes. The Luftwaffe flew 337 sorties against PQ18 and lost thirty-three torpedo-bombers, six long-range bombers and two reconnaissance aircraft. Meanwhile, *Avenger* joined the homeward-bound QP14, giving anti-submarine protection. Significantly, the convoy lost only one ship and the mine-sweeper *Leda*, whilst *Avenger* was in company, but two ships, a fleet oiler and the destroyer *Somali* were sunk after she had withdrawn.

German aircraft sank thirty-two ships from Russian convoys in 1942, but they sank no more until 1945 when one ship was lost. But this was not due to escort carriers, which did not sail regularly with the Russian convoys until early in 1944, because of the priority given, first to TORCH, and then to the Atlantic convoys. *Dasher* did join convoy JW53 in February 1943 but had to retire because of storm damage after two days.

Convoys to Russia were suspended in March 1943, and resumed in November. Escort carriers, each normally equipped with one 'composite' squadron of about twenty aircraft, Swordfish and Wildcats, returned to the Arctic when *Chaser* (816 Squadron, of eleven Swordfish and Wildcats each) went with JW57 in February 1944. On 4 March, on the way back south with RA57, one Swordfish piloted by Sub-Lt P.J. Beresford, with Sub-Lt D.F. Lang, and Leading Airman J. Beach as TAG, sighted a U-boat twenty-seven miles from the convoy and attacked with rockets, damaging her so badly that she was unable to dive. The destroyer *Onslaught* came up and sank U–472 by gunfire.

By now, the convoy was well into a pack of U-boats and sightings were frequent. Next day, another of 816's Swordfish (Sub-Lt J.F. Mason, Sub-Lt D. Street and Leading Airman D. Franklin) sighted U–366 on the surface. Mason used cloud cover intelligently until he was able to pounce on his prey and sink her with four direct rocket hits from a full salvo of eight. As if that was not enough, the next day, 6 March, dawned fair and full of promise. The dawn patrol struggled off the deck in a light wind and got six submarine contacts before breakfast. One Swordfish piloted by Sub-Lt L.E.B. Bennett (with Sub-Lt E.W. Horsfield and PO C.A. Vines) picked up U-973 by radar at twelve miles and sank her with a precisely aimed salvo of six rockets. Three more U-boats were sighted before midday and the same Swordfish with the same crew attacked and damaged another at dusk.

The Wildcats proved to be tremendous assets in the Arctic. They could be catapulted while the early Hurricanes could not, which meant they could be launched more quickly and in lighter wind conditions. Above all, they were robust and could be landed unharmed with an impact which would have broken a Seafire's back. *Chaser*'s Wildcats, Captain McLintock wrote, 'continued to take the most almighty clouts, and come up smiling. You apparently cannot ask too much of an American undercarriage beneath a good pilot.'

Activity (819 Squadron of three Swordfish and seven Wildcats) and *Tracker* (846 Squadron of twelve Avengers and seven Wildcats) went with the next convoys, JW58 and RA58 in March and April 1944.

Although neither carrier had had much time to work up and had never operated together before, they both co-operated in an 'all singing, all dancing' convoy escort in which their fighters shot down three Focke-Wulf Fw200Cs, two Junkers Ju88s and a Blohm and Voss Bv138C, while Avengers of 846 and the destroyer *Beagle* sank U–355 on 1 April, and two Swordfish with rockets of 819 and Avengers of 846 with depth-charges, and a Wildcat with machine-guns, together sank U-288. They also damaged U–362, U–673 and U–990.

But the prize for rapid and successful U-boat killing went to *Fencer* (842 Squadron of eleven Swordfish and nine Wildcats) who went with convoy RA59 in company with *Activity* in May 1944. Her Swordfish sank three U-boats with depth-charges in two days, two of them on the same day; U–277 on 1 May, and U–674 and U–959, both on the following day. In all cases the method of attack was by radar detection, careful and subtle approach, a swift arrival over the target and an accurate pattern of depth-charges.

The four British-built escort carriers, *Activity*, *Vindex*, *Nairana* and *Campania*, were especially successful in the Arctic. *Vindex*, with 825, a famous squadron (of twelve Swordfish and six Sea Hurricanes) went with *Striker* (824, twelve Swordfish and ten Wildcats) to escort JW59 and RA59A in August 1944. The fighters shot down three Blohm and Voss Bv138Cs. 825 Swordfish sank U–354 with depth-charges on 25 August, and co-operated with the destroyer *Keppel*, the escorts *Mermaid*, *Peacock* and *Loch Dunvegan* in sinking U–344 on the 24th and U–394 on 2 September 1944. Another successful squadron was 813 (twelve Swordfish and four Wildcats) embarked in *Campania*, which sank U–921 on 30 September 1944, and U–365 on 13 December.

For every U-boat sunk or badly damaged, a score had been driven off and prevented from attacking the convoy. An escort carrier, fully worked up and operational, was a marvellous demonstration of air power at sea, in which several kinds of Fleet Air Arm expertise in bombing, rocketing, strafing, fighter direction, radar presentation and interpretation, a captain's judgment, aircrews' skill and maintenance crews' devotion, all contributed to a fighting air group in miniature. The Swordfish was the most successful U-boat killing aircraft because it could be operated from the escort carrier's narrow decks by night. The Avenger was larger, stronger, and with its enclosed cockpits, very much more comfortable for the aircrews, but it was very seldom operated at night.

For everybody, ship's company and aircrew alike, the main enemy was the weather. Various carriers were 'arcticised' to various degrees, with steam heating of upper-deck fittings, guns, barriers and arrester-wire pulleys, but a major part of a carrier's effort was still the manual labour of constantly clearing the flight deck of packed snow and ice. For the aircrews, the Arctic weather made flying conditions very difficult and often dangerous. The aircraft flew for hours on end in darkness or in the opaque, grey semi-darkness of an Arctic winter day, through savage gales and blizzards which sometimes lasted for so long the soul despaired of ever seeing clear weather again. The bitter cold frustrated many otherwise successful attacking runs: depth-charge safety release clips were frozen solid at the crucial moment, rockets failed to ignite or launch successfully, guns misfired. To fire a gun at all, a TAG had to strip off several layers of gloves before he could feel the gun controls and he risked frost-bite every time his gun was in action.

After his patrols, when the pilot already had one eye on his fuel gauges, the pilot and observer had to search for their carriers which were very often hidden in fog banks, or freezing snow-storms. In an Arctic gale, the small carriers pitched and tossed like things demented, burying their bows deep in green sea and throwing tons of water in spray which froze as it swept across the flight deck. The crew of any aircraft which went into the sea had to be rescued in a matter of minutes, almost of seconds, before they froze to death. After hours of sitting in open Swordfish cockpits, it was common for aircrews to be so helplessly numb with cold that they had to be lifted bodily out of their seats. Some eighty aircraft were lost while escorting Russian convoys but very few to enemy action. Most were in flight deck-landing crashes, or in ditchings.

Young men forget, but one Swordfish observer remembered, after the war. 'Operational details are now so hazy that I can't even remember the number of the squadron in *Nairana* to which I belonged. On the other hand, images of storm and ice, the Aurora, the grey mid-winter noon, the bleakness of the great Arctic—all these I can savour as though they were recent happenings. I can still feel a moment of unalloyed happiness when, returning from a rather messy night operation against shipping in the Norwegian leads, *Nairana* parted company with her escorts off Shetland and made south to fly the squadron to Hatston, and take herself down to the Clyde. In the morning, I took a lonely constitutional on the flight deck. The sea was oily and the sky clear with winter brightness. A snow-covered Fair Isle was abeam and the absolute stillness was broken only by the musical and rhythmical throbbing of *Nairana*'s diesels. The war against Germany was drawing to its end and it might be that we would not be flying in anger again; but what really made the moment so charged with the feeling that I would hardly ever be as happy again was the ship moving through that crisp cold and brilliant sky, across the quiet sea.'

The Fairey Barracuda

As the war went on, a necessary love–hate relationship developed between the Fleet Air Arm and certain aircraft, usually expressed in faintly derisive verse about the machine's looks, handling qualities or armament—or the lack of them. But for the Fairey Barracuda, torpedo and dive-bomber, the compositions reached an intensity of outrage which almost amounted to a separate poetic genre. One of the aptest and wittiest, reputedly composed within the rival firm of Blackburn, found its way into the linebook of 827 Barracuda Squadron.

For why should the everlasting sky
 Be polluted and corrupted by
That product of a fertile brain
 That's aeronautically insane
That people call the Barracuda?
 Whose form is infinitely cruder
Than any form or scheme or plan
 As yet conceived by wit of man.

To see it flutter into space
 Would bring a blush upon the face
Of the most hardened Pharisee
 Within the aircraft industry.

But let us not decry
 This horror of the English sky.
Festoon its wings with fairy lights
 And wheel it out on Gala Nights,
And so dispel all hints, all rumour
 That Britain has no sense of humour.

One of the ironies of war, concealed by the need for wartime propaganda, was the contrast between the public praise of the Barracuda in the press and the private invective of the aircrews who had to fly it. In the fleet, it was generally held that the Barracuda had been designed by a committee and built of bits filched from the Forth Bridge. Like the bumble-bee, it was aerodynamically impossible, and certainly it had a disconcerting habit of failing to pull out of dives and, sometimes, even diving inexplicably into the sea from level flight, with the loss of its entire crew—as the sombre, black-edged 'In Memoriam' notices in the Barracuda squadron linebooks all too frequently showed.

The first Barracuda prototype flew in December 1940, but the type did not enter squadron service until January 1943, and first saw action with 810 Squadron in *Illustrious* off Salerno. The long delay betrayed the design problems of the Barracuda which showed themselves in the finished machine. The aircraft had been designed around a central 'chart-room' for the observer which actually had protruding 'bay windows' to give a better view. Also, to improve the downward vision, the wings were set high on the fuselage. This led to problems in designing a retractable undercarriage. Each solution to a problem bred further problems. The wings were so far from the deck that the tallest man could not reach them and special tackles and purchases had to be rigged to fold and unfold them. With its full array of torpedo or bombs, radar displays, aerials, dive brakes and flaps, undercarriage, lifeboat, strutted tailplane, tackles, toggles, purchases, footholds and intakes, the Fairey Barracuda did look like a 'flying monkey puzzle tree'. The aircraft *looked* awkward and a brute to fly, which it certainly was. The contrast between the Barracuda and its American equivalent, the Grumman Avenger—neat, robust and workmanlike, with a marked superiority in performance and load-carrying, and wings that could be folded and unfolded hydraulically from inside the aircraft itself—brought tears, if not of sympathy then certainly of rage, to the eyes of all who beheld it.

However, for all its short-comings, the Fairey Barracuda did win one outstanding battle honour, in the Fleet Air Arm's attacks on *Tirpitz*.

The Life and Death of the *Tirpitz*

It is very probable that Hitler, a land animal, never truly appreciated what a trump card he had in *Tirpitz*. She was the perfect example of the 'fleet in being'. All she had to do was, like Everest, just be there. Her presence, as Mr Churchill said, affected warship movements all over the world. She did not have to go to sea to be effective, indeed it was better she did not. At sea, she became one battleship, with a commander bedevilled by political caveats, a ship's company badly in need of sea time, an inadequate destroyer screen and patchy air cover. On 6 September 1943, *Tirpitz* sailed with *Scharnhorst* to carry out a short and unimportant bombardment of shore installations at Spitzbergen. It was the first and the last, indeed the only, time that *Tirpitz* fired her main armament in anger at a surface target. She then retreated to her northern fastness in Kaa Fjord, never sailed for another operation, and never saw the open sea again. But in harbour she had more effect upon the war than she ever had at sea. Her presence cast an ominous shadow across the way to Russia and over the Atlantic as a whole. Lying in her fortified lair, strung about with patrols and nets and booms, *Tirpitz* became, like Grendel's mother, a creature of legendary menace, accredited with supernatural powers. Shrouded in Arctic mists and clouds of rumour, she seemed to loom ever more monstrous and threatening the more the Allied naval staffs thought about her.

Such an ogre obviously could not be left unmolested. In fact, Hampdens of Bomber Command had attacked her as early as the autumn of 1940, when she was being fitted out at Wilhelmshaven, and she remained a prime target for all branches of Allied arms, from Norwegian resistance fighters to the Russian air force. But it was not until September 1943, when midget submarines penetrated Kaa Fjord and laid mines under her hull that *Tirpitz* was seriously damaged.

By March 1944, intelligence sources and photo-reconnaissance flights showed that *Tirpitz*'s repairs were nearly completed and she was getting ready for sea again. In that spring and the following summer the Fleet Air Arm mounted what can properly be called a campaign of strikes to try and disable *Tirpitz* where she lay.

The first strike, code-named TUNGSTEN, was the largest operation the Fleet Air Arm had ever mounted, involving two fleet and four escort carriers, 168 aircraft, and 204 aircrew (of whom 149 were reserve officers from five nations, fourteen regular RN officers and forty-one TAGS). The aircrews trained intensively for their task, with special scale models of the fjord, and a dummy range laid out at Loch Erriboll in Caithness, where the geography curiously resembled the upper reaches of Kaa Fjord.

The battleships of the Home Fleet, *Duke of York* flying the flag of the Commander-in-Chief, Admiral Sir Bruce Fraser, and *Anson* flying the flag of the Flag Officer Second in Command, Vice Admiral Sir Henry Moore, sailed from Scapa Flow, ostensibly to cover convoy JA58 to Russia, on 30 March 1944, with *Victorious*, the cruiser *Belfast* and destroyer escort. A second force, under Rear Admiral Bisset (Rear Admiral Escort Carriers), was already at sea, with Bisset flying his flag in *Royalist, Searcher, Emperor, Pursuer, Furious* and *Fencer*, the cruisers *Jamaica* and *Sheffield*, destroyer escort and oilers.

Victorious had two fighter squadrons, 1834 and 1836, with fourteen Corsairs each. Her Barracuda squadrons, with twenty-one aircraft, 829 and 831, formed No. 52 TBR Wing, but in order that the Wings which had trained together should fly on the strike together, she and *Furious* (No. 8 TBR Wing) exchanged one Barracuda squadron each. So *Victorious* actually sailed with her own 829 and *Furious*'s 827. Her total of forty-nine aircraft embarked was more than *Victorious*'s design capacity and some of the aircraft had to be permanently parked on deck.

Furious had her own 830 Barracuda Squadron and 831 from *Victori-*

ous, with twenty-one Barracudas, and two fighter squadrons 801 and 880, with eighteen Seafires. *Emperor* had two fighter squadrons 800 and 804, with twenty Hellcats, *Pursuer* and *Searcher* each had two fighter squadrons, 881 and 896, and 882 and 898 respectively, with twenty Wildcat Vs. *Fencer* had 842 composite anti-submarine squadron, with twelve Swordfish and eight Wildcat IVs.

The two forces met off the Norwegian coast on 2 April. The weather was fine, the ships had been undetected, radio traffic back in Scapa having been artificially maintained to conceal the fleet's absence. The crews had all been briefed and the Barracudas all bombed-up according to the complicated programme, some being armed with 1600-lb armour-piercing bombs, others with three 500-lb semi-armour-piercing bombs each, still others with three 500-lb medium-case anti-personnel bombs each, and a fourth with two 600-lb A/S bombs each. At 4.20 p.m., the Commander-in-Chief in *Duke of York*, with two destroyers, moved off to the north-west, while Admiral Moore led the TUNGSTEN force eastwards to the flying off position, which was about 120 miles west of the mouth of Alten Fjord.

At 4.16 a.m. the following day, on a perfect Arctic morning of clear, cold sky, a hard, blue horizon and visibility almost unlimited, the first Corsair from 1834 Squadron of the ten in the strike escort roared off *Victorious*'s flight deck. It was followed eight minutes later by the twelve Barracudas of 827, and from *Furious* the nine of 830. No. 8 TBR Wing being led by Lt-Cdr R. Baker-Faulkner. Twenty Wildcats of 881 and 882 took off from *Searcher* and *Pursuer*; and ten Hellcats of 800 from *Emperor*. Seafires from *Furious* and Wildcats from *Fencer* provided protective CAPs over the ships themselves.

The whole strike, of twenty-one Barracudas, and forty fighters, formed up correctly and took departure, flying south-east for Kaa Fjord, at 4.37 a.m. The Fleet Air Arm, with this formation of sixty-one aircraft, and another similar one yet to be launched, had come a long way from the last time it had operated over Norway in 1940—a point not lost on Rear Admiral Bisset, watching from *Royalist*'s bridge. 'It was a grand sight, with the sun just risen,' he wrote in his report, 'to see this well balanced striking force of about twenty Barracudas and forty-five fighters departing at very low level between the two forces of surface ships, and good to know that a similar sized force would be leaving in about an hour's time. It was especially heartening to an ex-carrier Captain accustomed for several years, to be very short of aircraft (especially fighters), and made one wonder "what might have been", if the Fleet Air Arm had been adequately supplied with aircraft in the early days of the War!' What might have been, indeed.

The strike flew low over the sea until they were some twenty-five

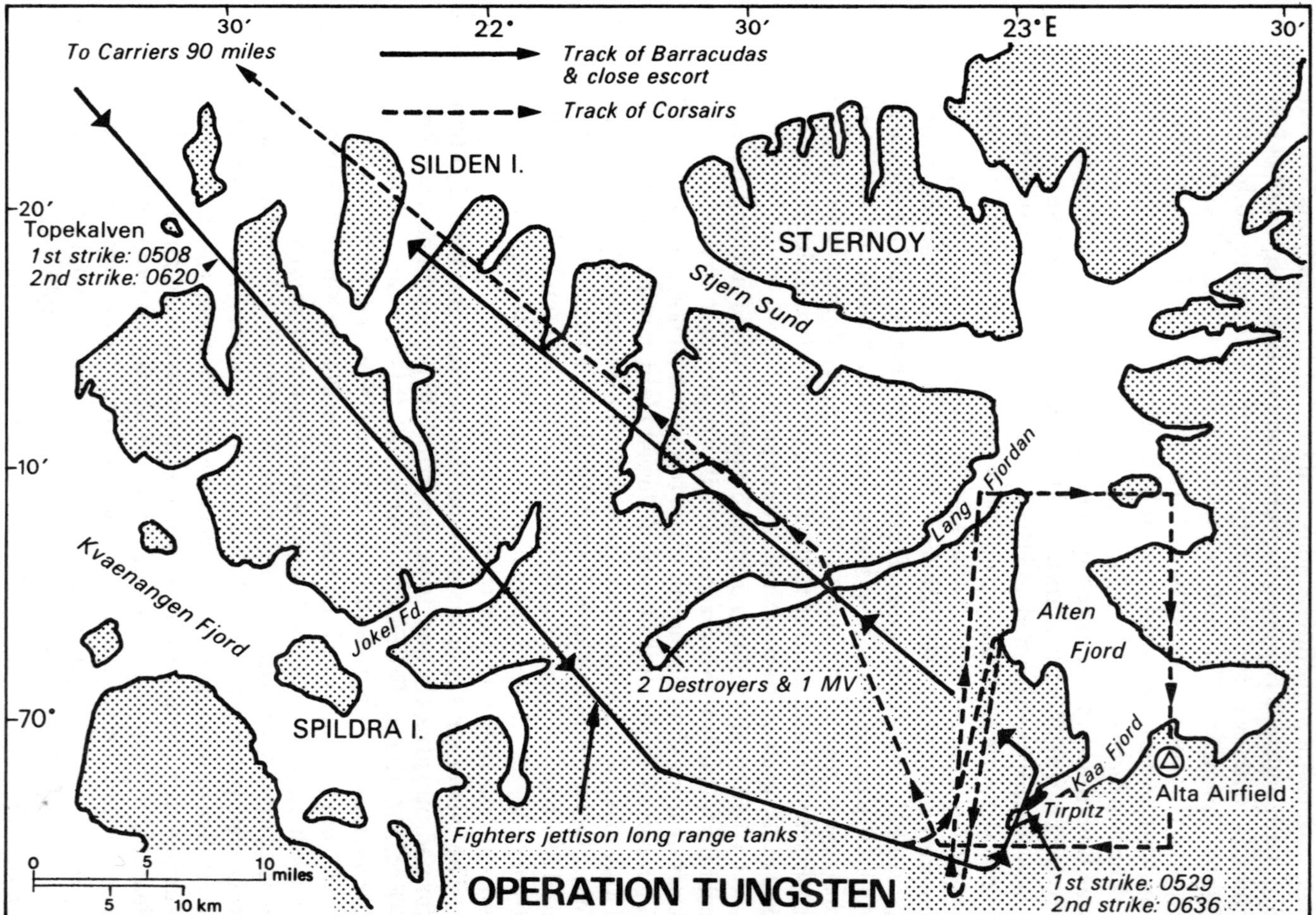

OPERATION TUNGSTEN

miles from the Norwegian coast, when they began to climb to 10,000 feet. They crossed the coast at about 5.10 a.m., flying south and east to approach *Tirpitz* from the south, intending to bomb her in two columns, along her fore-and-aft line (it had been seen that errors of range were greater than errors of line). They were undetected and unopposed.

Meanwhile, *Tirpitz* herself was weighing anchor, to proceed into Alten Fjord for a day's trials. The mood on board was hopeful. The damage of September had been repaired, the previous fortnight of trials in March had gone fairly well, and morale on board was buoyant, after the interminable months of looking out at nothing but frozen, hostile scenery. But, as the second anchor was coming in, smoke was seen wisping from the smoke-generators ashore. A report from shore warned of thirty-two aircraft heading south, range forty miles, and at 5.28, when only a faint smoke screen had been laid, the first aircraft was in sight, diving steeply on *Tirpitz* from the south. *Tirpitz*'s crew hurried to their action stations and to man their guns but not all of them were closed up before the first bomb arrived. While the Corsairs kept guard at 10,000 feet above, the first Barracuda plunged down through *Tirpitz*'s anti-aircraft barrage at 5.29 a.m. Sub-Lt D. Sheppard, RNCVR, watching from one Corsair, later said 'The first strike caught the Germans with their trousers down. We stooged around in high cover at 12,000 feet, I counted hit after hit, a dozen perhaps, or more. The *Tirpitz*? It was smoking and burning . . . just taking it.'

There was no air opposition of any kind, though there was some flak from the shore batteries, and the Barracuda crews could see their objective loud and clear. 'We picked up our target well,' said Lt K. Gibney in *Send Her Victorious*, an observer and Commanding Officer of 827, 'and saw a great black swastika painted on the fore-castle of the *Tirpitz*. We had one of the big 'cookies' aboard, and our wings lifted as it went down. I watched it hurtling down towards the *Tirpitz*. I couldn't see it hit, but the tail gunner of the following plane said it exploded in a cloud of smoke and flame.'

Tirpitz was unable to avoid the attack because of the nets around her and the land close to, and so, as the Corsair pilot said, she just had to take it. *Tirpitz* was hit by two 1600 pounders, a 600 lb, and three 500-lb bombs for certain, another two 1600 lb and one 500 lb probable. Meanwhile, the Hellcats and Wildcats of the escort flew low over the battleship, machine-gunning her gun positions and superstructure. In one minute the strike was over. Serious damage had been done internally, large parts of the upperdeck wrecked, many men killed or wounded. The trials fixed for that day would not now take place. (Actually the damage might have been worse. The heavy armour-piercing bombs were designed to be dropped from above 3000 feet for maximum effect but some of the crew in their enthusiasm had come down lower.)

Meanwhile, the second strike of nineteen Barracudas (one failed to start, a second crashed in the sea after take-off, killing the crew) of 829, from *Victorious* and 831 from *Furious*, led by Lt-Cdr V. Rance, No. 52 TBR Wing leader of *Victorious*, were just taking off. Escorted by ten Corsairs from 1836, nineteen Wildcats from 896 and 898, and ten Hellcats from 804, the second strike could see the smoke rising from *Tirpitz* from forty miles away. They attacked at 6.35 a.m. and in two minutes scored one 1600-lb and one 500-lb hit for certain and two probable 500-lb hits, while the fighters again strafed *Tirpitz*'s upper deck.

The last of the second strike had landed on by 7.58 a.m. and the force headed for Scapa. Two Barracudas, one from each strike, had been shot down and one damaged Hellcat had been ditched. Losses were the crews of three Barracudas—nine men. Behind them, the strikes left *Tirpitz* with the serious casualties of 122 killed and 316 wounded, and morale once more at a new nadir. Captain Meyer, of *Tirpitz*, who was wounded himself, looked along the bomb-blasted, bullet-marked and bloody upper deck and saw in the shambles of the superstructure, the ruin of six months' work.

Meanwhile, to the victor the spoils. The TUNGSTEN force entered Scapa on 6 April in triumph. In *Victorious* for the operation had been the Reuters man, Desmond Tighe, a faithful chronicler of the naval air war, whose brother was in the Fleet Air Arm. 'The carriers were welcomed home with full honours,' his despatch was quoted in *Send Her Victorious*. 'As we passed each ship the officers and ratings lined the quarterdecks, took off their caps and cheered, the sound reverberating across the blue waters. It was an inspiring sight, and, standing on the Admiral's bridge of the carrier, I felt very proud of being in such a ship.'

However, they had only scotched the snake, not killed it. Soon it had to be attempted all over again. A repeat of TUNGSTEN, code-named PLANET, for 26 April, was cancelled because of bad weather. On 15 May a strike (code-named BRAWN) was actually launched from *Victorious* and *Furious* but met 10/10ths cloud at 1000 feet over the Norwegian coast and had to return without making an attack.

The next strike actually to reach *Tirpitz* was MASCOT, on 17 July 1944. Forty-four Barracudas from No. 8 TBR Wing (827 and 830 Squadrons from *Formidable*) and No. 9 TBR Wing (820 and 826 Squadrons from *Indefatigable*) were escorted by eighteen of *Formidable*'s Corsairs (1841 Squadron), eighteen Hellcats from *Furious* (1840 Squadron) and twelve Fireflies of 1770 Squadron from *Indefatigable* (the Fairey Firefly two-seater fighter-bomber making here its operational debut). The Germans had learned from TUNGSTEN and set up an observation post on a nearby mountainside, to give advanced warning and to direct anti-aircraft fire including the main 15-inch guns. *Tirpitz* had fifteen minutes' warning and although there was still no German fighter opposition when the Barracudas attacked at 2.20 a.m., the flak was very heavy and *Tirpitz* herself was almost entirely obscured by smoke. One near-miss was felt in *Tirpitz* a minute later, but nothing else. One Corsair and one Barracuda were lost.

By now, Admiral Moore was convinced that the Barracudas were too slow to get from the coastline where they were reported, and over *Tirpitz* before the smoke had covered her. But the Admiralty felt that a series of repeated blows over a short period of forty-eight hours would wear down the enemy's defences and, a practical point, exhaust their immediate supplies of smoke-making canisters. So, after *Tirpitz* had made her last trip to sea, for exercises in Alten Fjord with the 4th Destroyer Flotilla on 31 July and 1 August 1944, the Fleet Air Arm launched four more strikes against her, code-named GOODWOOD I, II, III and IV, beginning with the first two on 22 August 1944.

Five carriers, *Indefatigable*, *Formidable*, *Furious*, *Nabob* and *Trumpeter* took part. The strike and escort consisted of thirty-two Barracudas (of 820, 826, 827, 828 and 830 Squadrons), eleven Fireflies of 1770, eight Seafires of 887, twenty-four Corsairs of 1841 and 1842, and nine Hellcats of 1840. The Barracudas and Corsairs did not reach the target but the Hellcats bombed through holes in the smoke clouds, without getting a hit. Five enemy aircraft were destroyed, for the loss of one Barracuda, one Hellcat and a Seafire. The same evening six Hellcats from *Indefatigable* escorted by eight Fireflies made a second attack with no hits. While the force was withdrawing that evening, U–354 torpedoed and sank the escort *Bickerton* and hit *Nabob* in the stern. Although a thirty-foot hole had been blown in her hull and she was well down by the stern, *Nabob* returned home safely, even flying off aircraft on the way.

Two days later, on the afternoon of the 24th, *Indefatigable*, *Formidable* and *Furious* mounted the heaviest attack of all: seventy-seven aircraft, with thirty-three Barracudas armed each with a 1600-lb bomb, ten Hellcats with 500-lb bombs, and five Corsairs with 1000-lb bombs, escorted by nineteen Corsairs and ten Fireflies. This time they achieved two hits. One, by a 500-lb bomb on the top of 'B' turret, was unimportant but the other, by a 1600 pounder should have put paid to *Tirpitz*. It penetrated the main armoured deck and went down through eight decks to lodge deep in the bowels of the ship where it *failed to explode*. The Germans afterwards found it had less than half its proper amount of explosive. Two Hellcats and four Corsairs were lost.

The last strike in the GOODWOOD series was delivered from *Formid-*

able and *Indefatigable* on 29 August. The bombing force was twenty-six Barracudas, two Corsairs and seven Hellcats, escorted by fifteen Corsairs and ten Fireflies. All the bombs were dropped, with several near-misses but no hits. A Corsair and a Firefly were lost.

The truth was that a big ship needed a big weapon, bigger than the Fleet Air Arm could properly deliver. It was delivered on 15 September by twenty-seven Lancasters of No. 9 and 617 Squadrons, flying from Yagodnik in Russia. They dropped seventy-two mines and sixteen 12,000-lb Tallboy bombs, one of which hit *Tirpitz* in the bows right forward, just before she disappeared in the defensive smoke. The Germans decided that they could not repair *Tirpitz* and moved her to Tromso where she was in range of bombers from the United Kingdom. On 29 October twelve Lancasters got a near-miss which damaged *Tirpitz*'s port propeller shaft. Finally, on 12 November 1944, thirty-two Lancasters of 9 and 617 Squadrons gave *Tirpitz* the killing blow, dropping twenty-nine Tallboys, with two direct hits and one near-miss close to on the port side. After extensive flooding, followed by a major explosion in the after magazine, *Tirpitz* capsized, with great loss of life. She had done less and achieved more than any other battleship in the Second World War.

After GOODWOOD, *Furious* was paid off for the last time. A venerable ship, she had done more than her bit and it was high time she was honourably retired. The other Home Fleet carriers—the fleet carriers until they left for the Far East, and the escort carriers (twelve of them at various times) until the end of the war—carried out more than thirty offensive operations in Norwegian waters between April 1944 and May 1945. Coastal shipping was bombed and strafed, mines laid, factories, harbours, oil storages, barracks were attacked; fighters covered cruisers and destroyers attacking enemy convoys. The last Fleet Air Arm strike in the West was made on 4 May 1945 at Kilbotn, near Harstadt. A strike of forty-four aircraft, Avengers from 846 in *Trumpeter* and 853 from *Queen*, with Wildcats from *Searcher*'s 882, attacked and sank a submarine depot ship and U–711 alongside her.

The Wrens

By September 1944 the strength of the WRNS had built up from *nil* in June 1939 to its wartime peak of about 75,000 officers and ratings, filling ninety different categories of duty and fifty different branches. There were, of course, some reactionary males who always looked upon the Wrens as a monstrous regiment of women, like the grizzled old Chief Cook who said that 'of all the 'orrible things this 'orrible war has done, these 'orrible women are the 'orriblest'. But anybody with any sense at all could see, as the war progressed, that there were large areas of life and duties in the Navy and in the Fleet Air Arm which simply would not be able to function at all without the contribution of the Wrens. By the end of the war there were very few aspects of the Service which the Wrens had not stamped with their own individual flavour.

The first categories of Wrens, in October 1939, included secretaries, cypherers, coders, clerks, accountants, typists, telephone operators, signallers, motor transport drivers, and cooks (although King George VI visited Portsmouth naval barracks in December and said he thought galley work was too rough for women). By 1940, there were Wrens serving in virtually every naval establishment and base in the United Kingdom, from Cornwall up to the Orkneys and from Kent across to Northern Ireland. In 1941, the Wrens went abroad. The first draft, of twelve cypher officers and ten Chief Wren special operators, were all killed when the ss *Aguila* was torpedoed and sunk en route to Gibraltar. This was a shock the Navy felt like the loss of a major war vessel. 'I knew every officer personally and the team was carefully chosen from volunteers,' wrote the Director WRNS at the time, Dame Vera Laughton Mathews. 'It would be impossible to picture a finer company—we sent our best.' However, undeterred, Wrens were soon serving all over the world, in Egypt, North Africa, Washington, the Persian Gulf, Ceylon, India and Australia.

In 1942, the categories of Wrens' duties had broadened to include radar operators, cinema operators, gunnery dome operators, recruiters, censors, submarine attack teacher operators, meteorological data plotters, operations room assistants, bomb range makers, vision testers, AA target operators, tailoresses, routeing officers, intelligence officers, orthoptists, and boats' crews. Wrens worked the decoding machine for ULTRA.

For the Fleet Air Arm, Wrens packed parachutes and looked after safety equipment, maintained radios, engines, air frames and torpedoes, interpreted photo-reconnaissances, instructed in escape and evasion techniques, and pedalled furiously around Yeovilton airfield with walkie-talkie sets for 'makee-learn' Fighter Direction Officers. They brought a touch of colour and an element of intuitive feminine logic into an otherwise austere wartime scene. Generally they had a higher boredom threshold than men, were meticulous about detail, and liked to work a project through to its end.

Many a romance between Wrens and aircrew flourished on the bleak windswept tundras of Scottish airfields, in draughty hangars on the Welsh coast, on steamy verandahs in Delhi, in the tropically scented nights of Kandy, in the breezy heights of the control tower at

Yeovilton. Squadrons gladly put on their Number One uniforms and best drinking boots to celebrate marriages between aircrew and Wrens, in all manner of places from Bournemouth to Boston, Bermuda to Bondi, Bombay to Brawdy. Some of the letters and occasional poems which Wrens wrote have such keen perceptions as can still touch the heart. One Third Officer WRNS, Olivia Fitzroy, serving at an air station in Ceylon, endured watching the pilot she hoped to marry killed in a flying accident. A poem she wrote in a Fleet Poetry Broadsheet published in Colombo in 1945 contains a truthful observation about the Fleet Air Arm, wrapped in authentic period language.

FLEET FIGHTER

'Good show!' he said, leaned back his head and laughed.
'They're wizard types!' he said, and held his beer
Steadily, looked at it and gulped it down
Out of its jam-jar, took a cigarette
And blew a neat smoke ring into the air.
'After this morning's prang I've got the twitch:
I thought I'd had it in that teased-out kite.'
His eyes were blue, and older than his face,
His single stripe had known a lonely war,
But all his talk and movements showed his age.
His whole life was the air and his machine,
He had no thought but of the latest 'mod',
His jargon was of aircraft or of beer.
'And what will you do afterwards?' I said,
Then saw his puzzled face, and caught my breath.
There was no afterwards for him, but death.

The East Indies

For much of 1943, in spite of Admiral Somerville's impatience for action, his theatre of the Indian Ocean remained in naval terms comparatively quiet. The escort carrier *Battler* with the composite 834 Squadron (twelve Swordfish and six Seafires) arrived in October 1943 for convoy defence duties and for the rest of the year she was the only Royal Navy carrier operating east of Suez. On 27 January 1944 Admiral Power arrived in Ceylon with three capital ships and the veteran fleet carrier *Illustrious* and the real rennaissance of the Eastern Fleet properly began.

Illustrious had two fighter squadrons, 1830 and 1833, which were among the first to be equipped with the big, fast, American fleet fighter, the Chance Vought F4U Corsair. The Corsair was a fearsome-looking machine which, with its crooked inverted 'seagull wing' and its general air of barely controlled menace, inspired as much fear in the hearts of those who were to fly it as in the enemy. One pilot of 1833, Lt N. Hanson, RNVR, who later commanded the squadron, first saw the 'monsters' they were to fly in a hangar at Quonset Point. 'Really, they looked killers to me,' he said in *The Forgotten Fleet*. 'They were the most dangerous bloody looking things I had ever seen. I'm not ashamed to admit it, that night I made a will.'

With an all-up take-off weight of 11,000 lbs, speed of over 400 m.p.h., range of 1100 miles, and an armament of two 0.50 machine-guns in each wing, the Corsair was a formidable fighting machine of an advanced design, with a hydraulic system which spread and folded the wings, lowered and retracted arrester hook and undercarriage, loaded and cocked the guns and generally did 'everything but cheer when you got a hit'. The earliest Corsairs had a built-in oleo 'bounce' in their undercarriages which caused so many deck-landing accidents that the US Navy temporarily discarded them from fleet work. However, the marque was successfully operated ashore by the US Marines, and by the Fleet Air Arm afloat, after the undercarriage had been rectified. The exceptionally long engine cowling gave very poor visibility forward and the aircraft had to approach a flight deck in a controlled but steep port hand turn, so that the pilot could see the batsman from over the port side engine cowling. For one pilot, Major Hay, Royal Marines, 'by far the most healthy improvement [in the Corsair's performance over British makes] was its endurance, with about five hours' worth of fuel in your tanks, you didn't have the agony of wondering whether or not you would make it back to the carrier.' *Illustrious* also had a Barracuda squadron, 847, which suffered the same trials and vicissitudes as all Barracuda squadrons.

At first, the repair carrier *Unicorn* was used as an operational flight deck, but, as there was also a shortage of destroyers in 1944, there being not enough to escort convoys as well as the fleet and the convoys taking priority, there was also a shortage of operational experience. It was not until April, with the arrival of the US fleet carrier *Saratoga* and three US destroyers, that Somerville was able to undertake a strike at Sabang, on a small island with a harbour and other facilities, off the north point of Sumatra.

The operation was carried out at the request of Admiral King, to put pressure on the Japanese in South-East Asia, while the landings at Hollandia were carried out. Eastern Fleet operations generally were a kind of muted accompaniment, in a minor key, to the greater undertakings in the Pacific. Aircraft from *Saratoga* and *Illustrious* attacked the harbour and oil storage tanks, and *Saratoga* was seen to be in a class above *Illustrious* in the theory and practice of operating carrier aircraft. The Fleet Air Arm clearly had some way to go before its aircrews could take their place in the main war to the East.

Early on Friday, 14 April 1944, Lt-Cdr 'Dickie' Cork, DSC, *Illustrious*'s Corsair fighter wing leader and one of the best fleet fighter pilots of the war, was killed in a needless accident at China Bay, near Trincomalee. Cork was approaching to land when another Corsair, piloted by a young pilot about to do his practice deck landings on *Illustrious*, was just about to take off. Cork flew round again but on his second pass, for some reason he ignored the danger light and landed. The young pilot could have turned off the runway but instead, stayed where he was, blinking his navigation lights on and off. The Corsairs crashed head on, burst into flames, and both pilots were killed.

In May, when *Saratoga* returned to the Pacific, the chance was taken to carry out a strike against the Wonokrono oil refinery and the harbour at Sourabaya in Java. It was a long-range operation, involving a passage of some 7000 miles, the fleet refuelling at Exmouth Gulf on the west coast of Australia. In June, again in response to a request from Admiral King, *Illustrious*'s aircraft struck at Port Blair in the Andaman Islands, and in July, after *Victorious* and *Indomitable* had joined the fleet, Admiral Somerville took the considerable force (for the Eastern Fleet) of two battleships, a battle-cruiser, the two carriers *Illustrious* and *Victorious*, seven cruisers and ten destroyers to sea for another attack on Sabang. *Victorious* also had two Corsair squadrons 1834 and 1836, of fourteen aircraft each, and 831, with twenty-one Barracudas. While the capital ships bombarded the harbour and destroyers engaged targets inside with guns and torpedoes, aircraft attacked airfields and shore facilities.

In August 1944 *Illustrious* went to South Africa for a short rest and refit, while *Victorious* and *Indomitable* attacked the cement works at Inda-roeng in Java and shipping at Emmahaven. *Indomitable*'s fighter squadrons, 1839 and 1844, were equipped with the Grumman F6F Hellcat, another excellent American fleet fighter with a top speed of 375 m.p.h., a range of 1100 miles, and three 0.50 Browning machine-guns in each wing. Lt-Cdr T.W. Harrington, No. 5 Fighter Wing Leader in *Indomitable*, thought very highly of the Hellcat as he wrote in *The Forgotten Fleet*: 'I recall their extreme reliability as well as their great flexibility for attacking anything from a high CAP right through the machinations of ground attack by gun and rocket; their usefulness as both a night-fighter without radar and a camera-carrying vehicle for photo-reconnaissance work. A further comforting feature of this splendid aircraft was that it was very ruggedly built and would absorb a great deal of enemy fire from astern.'

At Inda-roeng and Emmahaven, *Illustrious*'s 831 bombed both targets, while two of *Indomitable*'s Barracuda squadrons embarked, 815 attacked Emmahaven and 817 the cement works. In September, to coincide with the Pacific landings at Peleliu and Morotai, *Victorious* and *Indomitable*'s aircraft attacked Sigli in Sumatra and in October, timed to take place with the Leyte landings, the same two carriers' aircraft bombed and strafed targets on Car Nicobar and Nancowry islands.

It was, as everybody discovered, quite a different war out in the East, a war of heat and wide open spaces and long distances. It was no longer a matter of striking at the Italian battle-fleet and returning with all dispatch to Alexandria, or of bombing *Tirpitz* and being cheered back into Scapa three days later. In the East, operations took days which stretched into weeks and, eventually in the Pacific, were to turn into months. Because of the scale of distances, and the remoteness from the United Kingdom, carrier operations which would have been reported in the Press in large bold headlines in home waters, had only a column low down on the front page.

Old problems were intensified—aircraft reliability in heat and humidity, repair and replenishment of ships, intelligence which was scanty compared to the European theatre, maps which were few and inaccurate, escape and evasion for crashed aircrews which was quite a different proposition when the enemy was the Japanese.

In the East, air cover was as natural as breathing. Nobody of any consequence went anywhere without it. Any force of any size without a carrier in company felt naked and neglected. Yet, until 1943, there was virtually no supporting organisation ashore for naval air and very few facilities of any kind. RAF landing strips at Solur, Coimbatore and Cochin, in southern India, and at Colombo, Katukurunda, China Bay, the civil emergency landing ground at Puttalam, all in Ceylon, were converted to naval air stations. Aircraft from the United Kingdom or the United States were delivered to Cochin where they were assembled, air-tested and then ferried to maintenance units at Coimbatore. Katukurunda or Tambaran, where they were made ready for squadron service. Aircraft for operational requirements were held in reserve at Solur. China Bay was the immediate disembarkation airfield for fleet squadrons coming ashore from their carriers. By 1944, the Eastern Fleet's naval air organisation on shore could support thirty-four frontline squadrons and maintain some 400 operational aircraft.

In the carriers at sea, a favourite wartime paradox persisted: while the air groups complained that they were being called upon to fly on and on until they literally dropped, the ship's staff just as regularly exclaimed at the extreme youth and inexperience of the aircrews and asked, where have all the experienced pilots gone? Certainly Rear Admiral Clement Moody, the fleet's carrier admiral, had reservations about the fighting ability of his crews. Looking always to the East, to the big war in the Pacific, the aircrews had to practise their formation flying, forming up into large groups, maintaining fighter cover at several heights, bombing in wings, deploying by squadrons.

Each operation threw up a fresh set of problems. At Sabang in July, the form-up was slow and ragged, and it was too dark for the Corsairs to find their targets. Flak was nearly always light, but still picked off one or two aircraft from every operation. Enemy air opposition was very slight, and not many fighter pilots had the chance of combat. Sometimes, however, all went well, as in a good, crisp account of how it should be done (in *Send Her Victorious*) by Sub-Lt D. Heffer, RNZNVR of 1836, flying CAP on 25 July of Sabang: 'I was directed towards the enemy at 16.45 and sighted about five enemy aircraft. There was a large weather storm astern of the Fleet, but *Victorious* managed to direct me on to them. One of the enemy aircraft dived past me and after a short burst, I followed him down, hitting him on the port quarter with a long burst of fire. He was weaving, but flames were coming from his port wing, he disappeared into cloud and, following him, I came out dead on his tail at a range of about one hundred yards. After another long burst he went up in a sheet of flame.'

In 1944, *Battler* was joined by four more escort carriers, *Atheling* (889 and 890, with twenty Seafires), *Begum* (832 Squadron of twelve Avengers and four Wildcats), *Shah* (861 Squadron of twelve Avengers and six Wildcats) and *Ameer* (with twelve Avengers of 845). Admiral Somerville formed a carrier hunter/killer group around *Begum* and *Shah*, contrary to Admiralty advice that such groups should only be formed when there was a surplus of escorts and their operations should be directly linked to specific convoy movements. However, in August 1944, Avengers of 851 Squadron, from *Shah*, co-operated with the frigates of *Findhorn* and *Godavari* in sinking U–198.

On 22 November 1944, Admiral Sir Bruce Fraser hoisted his flag as Commander-in-Chief of the British Pacific Fleet, which eventually included four fleet carriers, *Illustrious, Victorious, Indomitable* and the new carrier *Indefatigable* (who had two Squadrons, 887 and 894, of Seafire IIIs, twelve Fireflies of 1770, eight Hellcats of 888 for photo-reconnaissance and 820 Squadron, of twenty-one Avengers). In December *Illustrious* and *Indomitable* began the first of a programme of strikes, code-named OUTFLANK, against Japanese oil refineries and storage tanks in Sumatra. The target was Pangkalan Brandan but

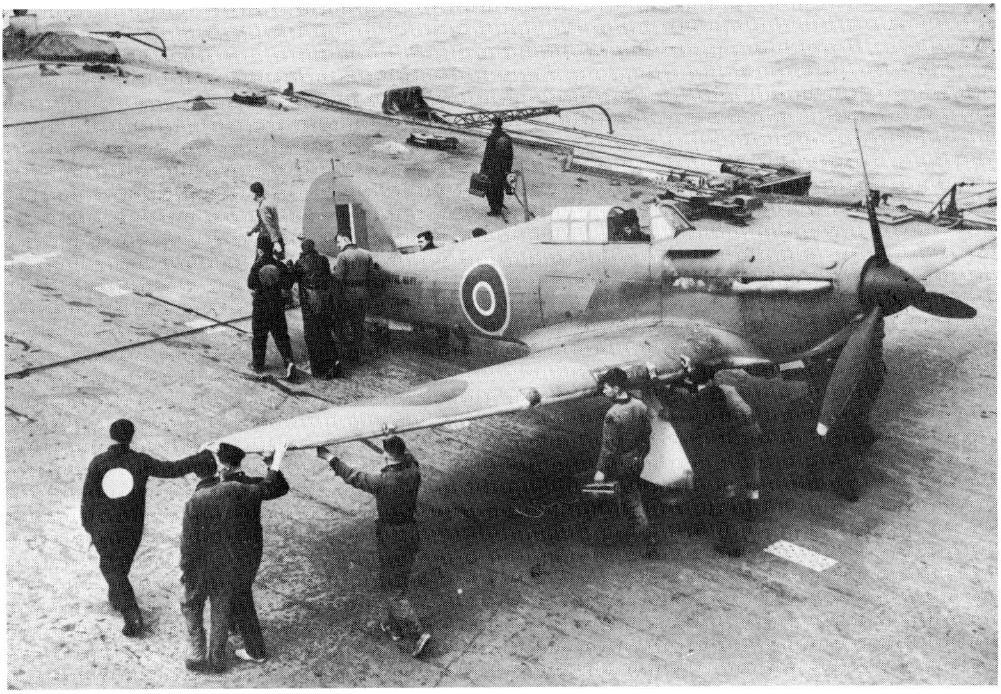

32 Sea Hurricane being spotted aft on the flight deck, with pilot's mate in the cockpit.

33 Scourge of the convoys: Focke-Wulf Fw200C Kondor long-distance bomber.

34 Reminders to those about to be catapulted.

35 Catapult Armed Merchant (CAM) ship catapulting her Sea Hurricane IA of the Merchant Service Fighter Unit.

36 'Try to imagine a ship cut off along the line of her deck and covered by a rectangular box with a flat overlapping lid. . . .' *Biter*, a typical escort carrier.

37 Swordfish being struck down on the lift in *Battler*.

38 Hellcat landing on *Ravager* during deck-landing training.

39 Avenger flying off *Fencer*, 1944.

40 'Ready aye ready'. Wildcat pilot, with book and umbrella, waiting for a scramble from *Fencer*.

41 The ultimate Stringbag: Swordfish Mark III fitted with Mark XI ASV radar scanner under nose, Rocket Assisted Take-Off Gear (RATOG) and Bristol Taurus XXX engine.

42 *Tracker*'s Avengers escorting a convoy to Gibraltar in 1944; *Biter* in the background.

43 Hellcats of 800 and 804 Squadrons in Arctic conditions on *Emperor*.

44 U-288 under rocket attack by Swordfish of 819 Squadron from *Activity*, in the Arctic Ocean, 3 April 1944.

45 To the tune 'Paper Doll':

I want a Barracuda I can call my own,
A plane the RAF can never steal:
And all those whiskered P/O Prunes
With their Mossies and Typhoons
Will have to fly in aircraft that are real.

As through the evening sky we slowly stagger:
Awaiting for the next poor chap to die
I'd rather have a Barra I can call my own
Than have an aircraft that can really fly!

46 Lt-Cdr R.S. Baker-Faulkner, of *Furious*, leader of first strike on *Tirpitz*, 3 April 1944.

47 Smoke screens beginning to drift across and hide *Tirpitz*.

48 Barracudas being bombed up for GOODWOOD in *Formidable*, August 1944.

49 In the ready room before the strike on Bödo, 26 April 1944.

50 Radio Wren: note the pre-1942 felt 'at.

51 *Victorious* in the Indian Ocean, 1944, with Corsairs and (aft) Barracudas on deck.

52 *Indefatigable* passing through the Suez Canal on her way out to the Far East in 1944.

53 Briefing on the refinery at Palembang. On the right, Lt-Cdr W.J. Mainprice, co of 854 Squadron Avengers of *Illustrious* who was lost with his crew when his Avenger struck a balloon cable over Soengei Gerong, 29 January 1945.

54 Attack on Palembang, January 1945.

55 *top* Hellcat peels off to port after a wave-off.

56 *bottom* Fairey Firefly I of 1770 Squadron on deck of *Indefatigable*.

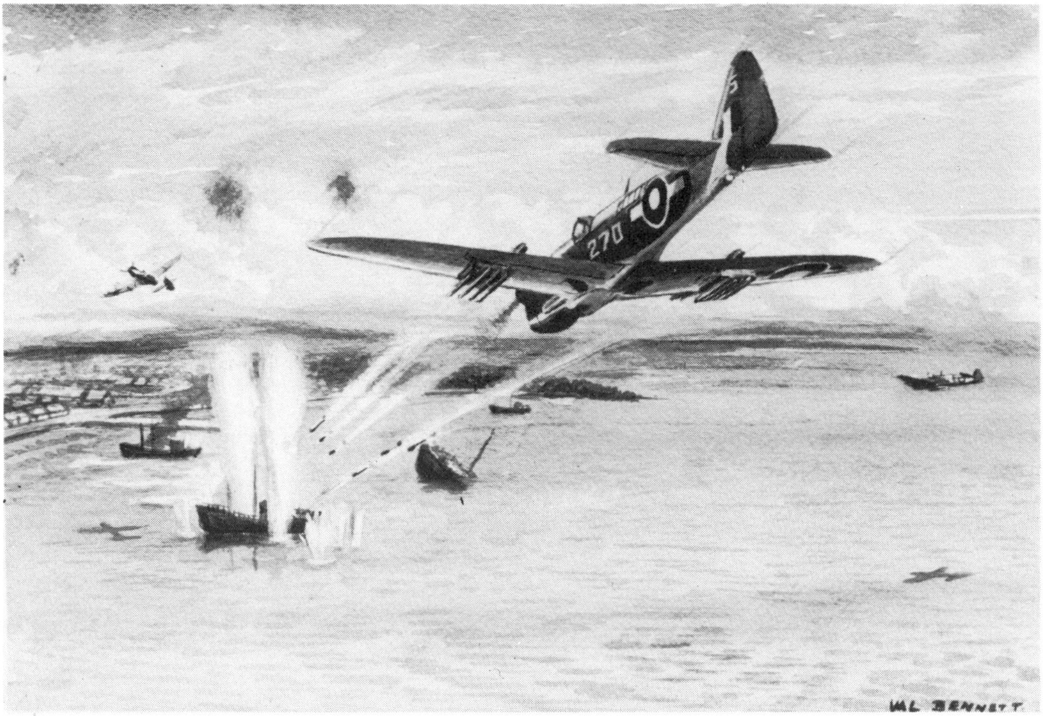

57 *top* Fireflies of 1770 strafing coastal shipping at Hirara, on Miyako, Sakishima Gunto, April 1945.

58 *bottom* Flight deck of *Formidable* after *kamikaze* hit of 4 May 1945.

59 Steward K. Mallender, of *Illustrious*, trundling a 500-lb bomb.

60 Massed Allied aircraft flying over the battleship *Duke of York*, Admiral Fraser's flagship, after the signing of the Japanese surrender, Tokyo Bay, 2 September 1945.

because of bad weather the strike diverted to the secondary target, the nearby port of Belawan Deli, where the town and harbour were bombed and strafed.

Early in the New Year, on 4 January 1945, *Indomitable, Victorious* and *Indefatigable* made a successful attack on Pangkalan Brandan. All four fleet carriers had landed their Barracuda squadrons and replaced them with Avengers (*Illustrious*, 854, *Indomitable*, 857, *Victorious*, 849, each with twenty-one Avengers). In spite of heavy and accurate flak over Pangkalan Soe Soe, the Avengers bombed the refinery, while the fighters, faced by their first serious opposition, shot down five Oscars, and destroyed another seven aircraft on the ground at Medan and Tanjong Poera.

The strike and escort returned in good humour, but, in fact, both operations had shown serious defects in technique. At Belawan Deli, radio discipline, never good, had deteriorated until the Strike Leader had been unable to make himself heard. Some fifty-four aircraft formed up afterwards, milling about in a state of indescribable confusion. Over Pangkalan Brandan, many fighters left their positions in close or high cover to engage in dog-fights all over the sky, and the Avenger crews complained that they were left unprotected as they deployed for their bombing runs. The fighter pilots replied that 'By attacking the enemy, we were *defending* the bombers.' Major Hay said (in *Send Her Victorious*) that 'it is a fairly natural reaction for any bomber pilot to feel that he is the only aircraft in the sky—seldom can you fly in neat, regular formation with the escort flying on your wingtip as a morale booster. The sky is never so empty as when you're flying and you appear to be all on your own with the enemy. The more so, if some of the escorting fighter Commanding Officers are inexperienced.'

Some of them were inexperienced, and some had R/T failures at critical moments and had to return to their carriers. Lt J.B. Edmundson of 1836 in his report, quoted in *The Forgotten Fleet*, did his best to disarm criticism. 'What always seems to happen is that someone sees a "bogey", makes a hasty report, and chases off, followed by everyone else who is anywhere near. The fault is, actually, that it's not every month of the year that you see a Zero in the Fleet Air Arm, so you can hardly blame a fighter pilot for making the most of his opportunities. I think the Avengers are a pleasure to escort, and the more we do it, the better protection they will get.' The Avenger crews had to be satisfied with that reassurance.

Despite all their efforts, the Fleet Air Arm had not provoked any real response from the Japanese, who were so hard-pressed in the Pacific that nothing short of a full-scale seaborne assault on Rangoon (and, conceivably, not even that) would have caused them to move major resources west into the Indian Ocean. As the war correspondent A.W. McWhinnie, cruelly but correctly wrote in the magazine *Illustrated* in February 1945, British naval strikes against the Japanese 'were of the tip and run variety, and that certain individual efforts amounted to little more than banging at the back door of the Japs and running away before the door was opened.' However, before the end of January 1945, the British Pacific Fleet carrier aircraft had completed two strikes which banged on the back door of the Japanese and stayed to punch on the nose whoever answered it.

6

1945 The Closing Stages

Palembang

There were many excellent reasons why the BPF should not attack Palembang. The targets lay inland, a long flight over enemy-held territory, and were very formidably defended by fighters and anti-aircraft guns. They were too difficult an undertaking for the BPF's aircrews at this stage in the fleet's existence. Oil refineries, in any case, were better attacked by heavy bombers. The Australian government had been promised that the BPF would arrive in Australia by the end of 1944 and if the fleet delayed to complete the OUTFLANK programme, this date could not be met.

Admiral Fraser replied to all these objections. He did not believe the Japanese were as strong as reported in the Indian Ocean, and he was sure the BPF's aircrews were up to the task. Oil refineries, with some vital vulnerable installations set in a vast insensitive area, were much better attacked by carrier aircraft's precision bombing than by area saturation by heavy bombers. It was unrealistic to expect the fleet in Australia by the end of 1944; the re-equipping with Avenger squadrons had taken longer than expected, but, even if the fleet had arrived, not enough essential elements of the fleet train had also arrived to enable the fleet to operate. Lastly, but most importantly, Admiral Fraser had already been to Pearl Harbour to meet Admiral Nimitz and to talk about the BPF's future. Fraser had so far impressed Nimitz with the capabilities of the BPF that Nimitz had asked if the fleet could strike at the Sumatran oil refineries on their way east to Australia. To refuse now, at the first time of asking, would be unthinkable. It was a matter of honour. So, as Admiral Fraser himself said, everybody at home gives in when the man on the spot insists. The Palembang strikes, which turned out to be the largest single operations ever mounted by the Fleet Air Arm during the war, were definitely on, weather permitting.

At least there was no argument about the strategic importance of the targets. Palembang, an important road and rail centre in south-east Sumatra, about 150 miles from the western coast, lay on the north bank of the Musi river, some forty miles from its mouth. Pladjoe (Royal

Dutch Shell), the largest refinery in the Far East, and Soengei Gerong (Standard Oil), the second largest, lay about five miles downstream of the town, on either side of the Komerine river where it enters the Musi. By January 1945, Japan was critically short of fuel and of tankers. The Palembang refineries, with Pangkalan Brandan to the north, were capable of supplying nearly two-thirds of Japan's requirement of aviation fuel. A blow at them would be a blow at the Japanese war effort indeed.

The aircrews trained as hard for Palembang as for any operation in the war, with intensive exercises, of up to a hundred aircraft, practising squadron and air group attacks using the coastline of Ceylon as a rehearsal target. There were also scale models of the refineries and the crews had the advice of former Dutch employees in the oil fields.

The fleet, designated Force 63, and virtually the British Pacific Fleet as first constituted, sailed from Trincomalee on 16 January 1945, to rendezvous with the oiling force. All four carriers, *Indomitable, Victorious, Illustrious* and *Indefatigable,* were present, with Rear Admiral Sir Philip Vian, Rear Admiral 1st Aircraft Carrier Squadron, flying his flag in *Indomitable.* Vian had followed the sound of the guns throughout the war and his war experience went back through the Normandy landings, to Salerno, to the Mediterranean convoy defence actions, to the *Altmark* Incident and the Norwegian campaign of 1940. An aggressive and successful leader, no better officer could have been chosen to lead the carriers in the new war in the Pacific which up to then had been almost entirely an American undertaking. Force 63 also included the battleship *King George* V, with Vice Admiral Sir Bernard Rawlings, who was to lead the BPF at sea, on board but not flying his flag; four cruisers; and two flotillas, of ten destroyers.

Bad weather at first delayed the strike but by dawn on 25 January, the force was some thirty-five miles off the western coast of Sumatra. The weather was clearing, the crews were ready to go, and so, at 6.15 a.m. Admiral Vian ordered the strike to be launched.

There were to be three strikes, the first at Pladjoe (code-named MERIDIAN ONE) the second at Soengei Gerong (MERIDIAN TWO) and a third to mop up the remains of both. The strikes at Pladjoe, and at Soengei Gerong, which followed on 29 January, were both on the same plan: a strong force of Avengers each armed with four 500-lb bombs, drawn from all four TBR squadrons, were escorted to and from the target by Corsairs and Hellcats, stacked at three levels, as high, middle and close cover. *Indefatigable*'s Fireflies also accompanied the strike, as part of the escort, but were also armed with eight 60-lb rocket projectiles each, to attack specified targets. Meanwhile two Corsair

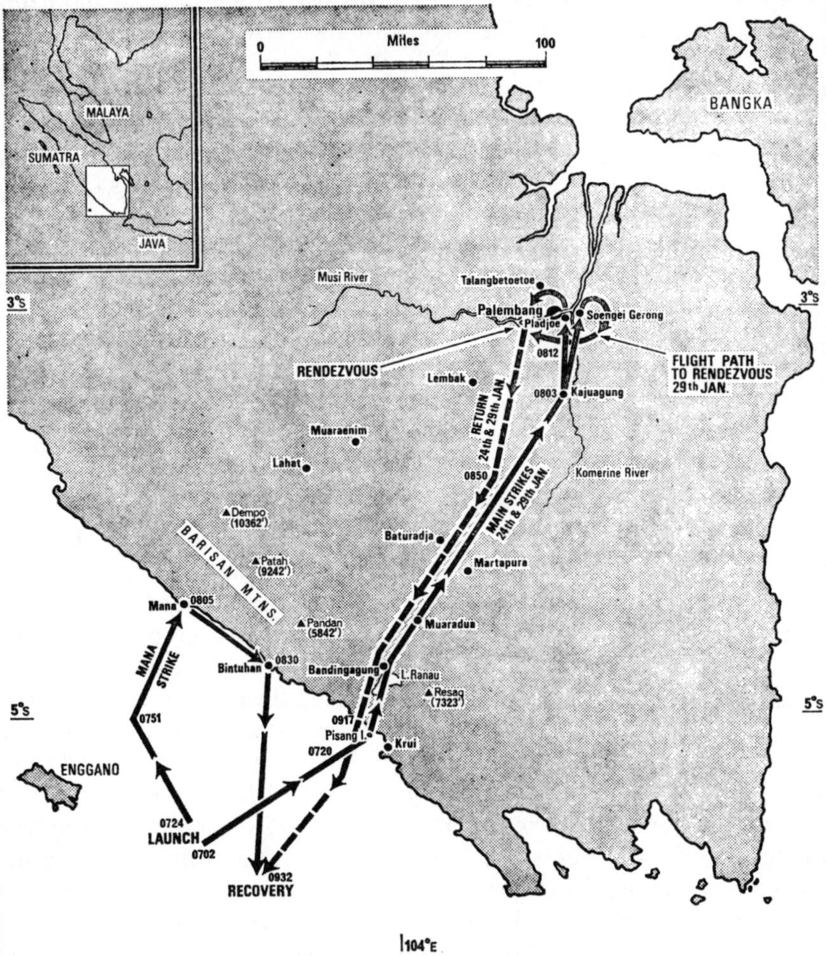

BPF STRIKES at SUMATRAN OIL REFINERIES, PALEMBANG, 24th and 29th January 1945

(Times are given for 24th. Times were similar on 29th)

groups, 'X-Ray' and 'Yoke', would carry out offensive sweeps (code-named RAMRODS) over the enemy fighter airfields (of which there were several within forty miles of the refineries) to keep Japanese heads down and prevent fighters taking off. The remaining Corsairs and Hellcats, with *Indefatigable*'s Seafires, provided CAP over Force 63. A subsidiary strike of five Avengers and four Corsairs attacked the coastal town of Mana. Hellcats flew photo-reconnaissance and leaflet dropping sorties. Two Walrus amphibians, 'Darby' and 'Joan', were ready in *Illustrious* for air-sea rescue duties. Altogether the force had some 244 aircraft embarked all of which were involved in some way.

The Strike Leader was Lt-Cdr W. Stuart, RNVR, Commanding Officer of *Indomitable*'s 857 Avenger Squadron and the Air Co-ordinator was Major Ronnie Hay, Royal Marines, of *Victorious*, whose flying experience went back to the beginning of the war and who was eventually one of the very few Fleet Air Arm pilots to serve throughout the war, and survive. The task of the Air Co-ordinator (a concept borrowed from the US Navy and tried, with success, for the first time in the OUTFLANK series of strikes) was to accompany the strike and direct the approach to the target and the withdrawal, Avenger squadron Commanding Officers remaining responsible for their squadron's individual attacks. As well as his co-ordinating duties, Major Hay took some excellent oblique-angle photographs of the bombing results, and he and his flight of four Corsairs also found time to shoot down three Tojos and an Oscar. His report, moreover, provided an excellently, dry and astringent commentary on the operation.

The first strike of eighty-seven Avengers, Corsairs and Hellcats took departure at 7.04 (nine minutes late, as Hay pointed out in his report), and 'for reasons unknown one Avenger from *Indefatigable* started to straggle. I have a photograph of this and have sent it to the squadron in the hope that it will teach them what not to do.' The strike was detected crossing the coastline and the first cry of 'Bandits' was soon heard, as the escort was engaged by some twelve or fifteen Japanese fighters. The Japanese pilots included some instructors and their standard of flying was generally high 'except for one laddie,' wrote Lt-Cdr Hanson, the Close Escort Leader, in *The Forgotten Fleet* 'who made an awful shambles of it. I saw loads of stuff coming over my wing, someone firing dead astern of us. I looked in my rear view mirror and there was this Oscar as big as a house. Everything happened in split seconds, I was going to shout "break" one way or another but before I could get the words out of my mouth or press the button the fellow disappeared and shot underneath me. Then, to my amazement, he appeared right in front of us, in absolute plain view. We all heaved back on our sticks and belted him. When you hit someone with point-fives something goes. Chunks flew off him and he fell off sideways in a sort of tail glide. He was only the length of a room away from us. We damned nearly ran into him.'

Meanwhile, through a growing barrage of heavy and light flak, so accurate for height, the crews could hear it deafeningly close to their cockpits, but not so accurate for bearing, the Avengers could see their targets – a great grey-green expanse of refinery, set in the bright green jungle, with the smoky traces of the town beyond, some small dusty white buildings, blue and grey storage tanks, a few striped with red, some attempts at camouflage, the line of roads, a scattering of trees and beyond, the oily glint of the river, gleaming like some great fluid muscle in the sunshine. On the perimeters of the refineries, the Avenger crews could see something quite unexpected: about thirty olive- or khaki-coloured barrage balloons, their sides patched and stained, at 2000 feet, and rising.

The balloons were a sure sign of the importance of the target and a most disturbing development for the Avengers. Lt-Cdr Mainprice, Commanding Officer of *Illustrious*'s 854 Squadron, was lost with his entire crew to the balloons on the second strike against Soengei Gerong. His wingman Sub-Lt R.S. Armstrong and his crew were also lost. 'It is a very unpleasant experience,' wrote Sub-Lt R.W. Halliday, an Avenger pilot of 854, in Peter C. Smith's *Task Force 57*, 'from the point of view of a pilot, to be flying along one moment with an aircraft flying perfectly normally beside you—and then suddenly the wing goes and the other plane spins down. We lost quite a few aircraft that day.'

The Fireflies were asked by the Strike Leader to strafe the balloons but never heard the request and dived down to attack their own targets. The Commanding Officer of 1770 Squadron was Major V.B.G. Cheesman, Royal Marines, who had been a survivor of *Cornwall* in 1942 and had a personal score to settle: 'Now we approach the target, and what a target, sitting there in the morning sunrise!' he wrote, in *The Forgotten Fleet*. '"Attack, attack, attack!" and in we go. Now don't rush it. Take careful and accurate aim. Remember how those bombs hit *Cornwall*. Don't muck it up at the last minute by being over-enthusiastic. Down we go, 300 knots, closing fast, sights on, hold it, hold it, fire! I hear my rockets go. Switch to cannon and a burst of cannon shells all on their way to the target, a tower of machinery known as a "cracking plant" which should be full of petrol. And by Jove, it is! An enormous explosion takes place and sheets of flame reach up into the sky.

'Watch out. A Jap fighter trying to get on my tail. Avoiding action and try to get on to his, but he's away now. The whole place is flame and smoke and chaos.'

'Throughout the attack,' wrote Hay in his report, later published in a supplement to the *London Gazette*, 'the enemy had just sufficient fighters to saturate the escort. Enemy pilots showed as much contempt for Japanese heavy AA as we did, and fights were raging all over the target area. It was almost funny to see the aircraft scrapping and all the while the AA bursting at all heights up to 15,000 feet. As far as I know no one was lost by this fire and very few damaged.

'The presence of three or four twin-engined aircraft seemed to indicate some air-ground control of AA or fighters. No noticeable difference was observed after three of these twin-engined aircraft had been shot down.'

The presence of so many fighters also showed that the RAMRODS had not managed to keep the enemy's head down. Flown off in the second range, they were just too late over the airstrips, to prevent Zekes, Tojos and Oscars taking off. The RAMRODS themselves lost several Corsairs from both strike days, to AA fire, or in combat, or in mid-air collision.

The Avengers, too, though once again feeling starved of fighter cover, defended themselves with tremendous courage and panache. The rear ball-turret gunners of several Avengers beat off successive attacking runs by the much faster Japanese fighters and succeeded in shooting down two of them. For one Avenger, flown by Lt E. Beeny of 857, everything went wrong. As he wrote in *The Forgotten Fleet* their radio was unserviceable so that they missed vital last-minute instructions about balloon evasion; their bombs hung up on the first run and they had to go round again; and then the TAG sighted a Japanese fighter closing from astern. After evading him for a time, the fighter attacked again and the TAG 'thinking at last he had him in his sights, pressed the button—and the gun jammed! Fortunately for us the Commanding Officer of one of the fighter escort squadrons had spotted us and the Tojo, and had given him a good burst and winged him. But the Jap wasn't quite finished yet. He next tried to ram us and was actually flying level with us when his tank caught fire and he rolled over and crashed just by his own aerodrome.'

The fighters were doing their best. 'Top cover was shouting the odds,' wrote Lt-Cdr Hanson in Richard Abrams' *F4U Corsair at War* 'and the battle was joined. The air became alive with warning shouts, orders to close up and all the natter that excitement generates. And then I could see them. Over to the North, pinpricks of black were hurtling downhill from a great height and con trails were streaming out across the unbelievably blue sky as our fighters pulled tight corners to get at them. The Avengers were now deploying for their bombing run and their line was lengthening. Christ! I could do with twenty Corsairs right now! Suddenly a Jap levelled out over the port side, going like smoke as he made his run to take the bombers on the beam.

'"Break left—Go!"

'The flight wheeled over on to its wingtips. We gave him a burst head-on and he ducked. Whether or not we hit him I don't know, but fire from twenty-four 0.5 guns is enough to make the bravest put his head down.'

After bombing, the Avengers made their way through more flak to a rendezvous north of the town and from there back to their ships—insufficiently escorted, in the Air Co-ordinator's opinion, by fighters. 'The Fireflies appeared to be the only aircraft there and they were unable to prevent several attacks by fighters taking place. There is no doubt more of the escort could have got there and discipline on this point must be tightened.'

This point was made even more forcibly by a Tojo pilot, Major Hideaki Imayama, of the Japanese 87th Air Regiment, who after the war gave a rare but graphic account (published in the *RAF Flying Review*) of the loss of a Fleet Air Arm crew, written by the opponent responsible. 'At 1500 feet two Avengers were flying southwards, their leader trailing smoke. Sitting ducks. I carefully turned in behind them, concentrating on the damaged Avenger which still had its bomb doors open. Probably its hydraulics had been damaged. Six hundred yards, five hundred yards. Suddenly its ball-turret gunner opened fire. Red tracer slipped past my Shoki (Japanese name for Tojo, meaning "Demon") but I held my fire. Two hundred yards. I could clearly see the gunner in the ball-turret. Now I was flying in the wash of my quarry and my aircraft was bouncing around like a mad thing. Steadying up the Shoki I fired at point-blank distance. The bullets from my four 13mm guns ripped into the Avenger, its green-house canopy bursting into fragments, like leaves in a gale. Flames seared back from the port wing roots and the Avenger rolled on to its back and then fell away into the jungle below.'

However, Major Hay concluded 'I think this has been one of the better strikes the Fleet Air Arm has ever accomplished.'

On the 29th, after the ships had refuelled, the second strike of forty-five Avengers, escorted by twenty Corsairs, sixteen Hellcats and ten Fireflies, with two more RAMRODS of twelve Corsairs each took off for Soengei Gerong. They were, as the Air Co-ordinator was quick to notice, late again but only by four minutes. At 7.33 a.m. Major Hay 'observed an aircrew being picked up by a destroyer: it seemed a little early in the day to start losing aircraft,' he commented, somewhat heartlessly.

Once again, the Tojos and the Oscars attacked and the guns opened fire. Again, the Avengers dived through the balloons and the flak barrage, dropped their bombs and made a rendezvous to collect themselves for the journey home. Sub-Lt Halliday, who had watched his Commanding Officer go down on the balloon cables, was as he said in *Task Force 57* himself 'hit on my way down to the target, but I didn't realise it until I pulled away, and saw three or four neat holes in my wing and flames coming out of the holes. I just burned steadily all the way back to the coast—I did not expect to make it as it lay ninety miles away and on the far side of a mountain range. I was flying very low down when I was hit and realised that I would have to gain altitude to clear these mountains before I could reach "friendly" territory. I had a lot of trouble: neither the wheels nor the flaps would lower, the hydraulics had gone, and the engine began playing up—it would only run, for some unknown reason, at a fantastically high rate of revs and every time I throttled back it threatened to stall and stop altogether. I just kept going as I was. I flew like this for half an hour with the wing blazing away like a torch and I cannot understand why it did not drop off miles back. Anyway, I reached the sea and to my intense relief saw a destroyer below me. I banked down past her, fired a Very light and pancaked close by her. She proved to be HMS *Whelp* and they had us all inboard in double quick time.'

Major Hay, meanwhile, was still observing events with a professional eye. 'About three minutes after the last aircraft bombed I finished photographing. I then climbed from 6000 feet to 10,000 in order to take vertical-line overlap photographs as the flak died down. I soon had to change my mind as a Tojo was coming for us. In shooting this one down, we descended to nought feet and, attracted by the gunfire, an Oscar came along, and by 0905 he, too, was dead.

'During this time the radio was giving me an interesting picture of a long stream of Avengers dribbling out of the target area to the rendezvous, thirty miles away. It was evident some of them could not find it. Nor could I from 7000 feet, even though I searched for some time. During this period there was quite a vicious air battle of which I can give no detail.

'. . . R/T discipline up till now had been 100% improved. But crossing the coast seemed to be the signal for complete radio chaos. Primarily the Avengers giving their damaged friends extracts from pilot's handling notes. It is about time everybody knew their emergency drill without having to talk about it.'

Halliday and his crew were among the luckier ones. Nearly all of the Avengers were badly battered during the raid and Halliday's was one of six that had to ditch near the fleet. In the two days' strikes, the fleet had lost six Avengers, eight Corsairs, a Hellcat and a Firefly to enemy action. Eleven aircraft had ditched and another fourteen were written off in deck-landing crashes, making a total loss of forty-one aircraft, from 378 sorties, a loss rate of over ten per cent. *Victorious* had lost eleven aircrew, seven from 849 Avengers. Ten of her nineteen Avengers were missing, ditched or unserviceable. *Illustrious* had lost ten aircrew, including Mainprice, her Avenger wing leader, and Lt A.W. Sutton, Senior Pilot of 1830 Corsairs.

But the results were worth it. At Pladjoe hits had been achieved on the crude oil distilleries and run-down tanks. About thirty per cent of the cracking and distillation units had been destroyed and there were hits in the main boiler and electric power house. At Soengei Gerong, photography had to cease about half-time because smoke was obscuring the target, but the cracking plant, coke stills, fractionating columns were all hit and boiler and electric power houses probably hit. Both refineries were put out of action for about a month and Soengei Gerong for nearly three months. Production had improved to about half normal by May 1945, but the refineries were not at full flow again for the rest of the war. Sixty-eight enemy aircraft had been destroyed, thirty in air combat, the rest on the ground.

Over the fleet, the Seafire CAPs had had mixed fortunes. On 25 January, the Seafires had suffered a sorry tale of mishaps, deck-landing crashes, smashed undercarriages, burst tyres and ditchings in the sea. That afternoon, the fighter cover for the fleet was taken over by the other carriers because, as Vian signalled, of *Indefatigable*'s slowness in operating Seafires. Captain Q.D. Graham, commanding *Indefatigable*, noted in his report that 'No. 24 Fighter Wing finished their day in a very low state of spirits'.

However, the Seafires redeemed themselves on the 29th, when Japanese aircraft approached the fleet several times during the day and Seafires shot down four of them, assisted by Hellcats of *Indomitable* and Corsairs from *Illustrious*. The Seafire squadrons were delighted and Captain Graham 'was glad to see that No. 24 Fighter Wing ended their day in a very high state of spirits'. But shells from the cruiser *Euryalus* had accidentally hit *Illustrious*'s flight deck and inflicted casualties of twelve killed and twenty-one wounded.

Special arrangements had been made to recover pilots forced down. The submarine *Tantalus* was on station off the Sumatran coast and a Walrus patrolled Lake Ranau, some miles inland, where crashed pilots had been briefed to make their way. Nine airmen were captured by the Japanese and taken to Changi Gaol in Singapore. In March, all nine were driven to the beach north of Changi where they were beheaded and their weighted bodies dropped in the sea. They were Sub-Lt J.K. Haberfield, RNZNVR, of *Indomitable*, Sub-Lt E.J. Baxter, RNZNVR, and Sub-Lt R.J.A. Shaw, RNVR, both of *Illustrious*, Lt K.M. Burrenston, RNVR, and Sub-Lt W.E.J. Lintern, RNVR, both of *Victorious*, with two unidentified sub-lieutenants and two petty officers. Three of the Japanese officers responsible for this war crime, the first committed by the Japanese against Fleet Air Arm personnel, committed suicide before they could be brought to justice.

Last Days in the Indian Ocean

After striking at Palembang, the ships of the British Pacific Fleet went on to Australia, leaving those staying behind in the Indian Ocean inevitably feeling 'second class citizens', a 2nd XI, left stranded in a backwater while the main tide had passed by. However, there was still plenty of action in the Indian Ocean until the end of the war. *Ameer*'s twenty-four Hellcats of 804 Squadron covered the assault landings at Cheduba and Ramree, on the coast of the Burmese Arakan in January 1945. *Ameer* was joined by *Empress* for photo-reconnaissance flights by 888 Squadron Hellcats in February and March. On 1 March, Hellcats from 804 'splashed' three Japanese aircraft—the first to be shot down by aircraft from British escort carriers. *Khedive* (twenty Hellcats of 808) and *Emperor* (888 PR Hellcats and eight Avengers of 851) sailed for more PR flights over Port Dickson and Port Swettenham and gave cover for fleet bombardments and anti-shipping sweeps, as well as shooting down Japanese aircraft over Sabang and Emmahaven.

By Operation DRACULA, the assault landing on Rangoon on 2 May 1945, the escort carriers had been formed into the 21st Aircraft Carrier Squadron under Cdre G.N. Oliver flying his broad pennant in the cruiser *Royalist*. *Hunter* (thirty Seafires of 807), *Stalker* (thirty Seafires of 809), *Khedive* (twenty-four Hellcats of 808) and *Emperor* (twenty-four Hellcats of 800) gave close cover to the assault forces, while *Empress* (twenty Hellcats of 804) and *Shah* (ten Avengers of 851, four Hellcats of 804) accompanied the fleet out at sea striking at the Andaman and Nicobar Islands, to keep the Japanese occupied and prevent them interfering with the landings at Rangoon.

The covering forces had no sooner returned to Trincomalee after DRACULA than intelligence reported that the Japanese were attempting to evacuate the garrison of the Andamans using the heavy cruiser *Haguro*. The fleet put to sea again, 851 Squadron's Avengers being transferred to *Emperor* because of defects on *Shah*'s catapult. On 15 May, south-east of the Andamans, an Avenger of 851, flown by the Commanding Officer, Lt-Cdr M.T. Fuller, RNVR, sighted and accurately reported *Haguro*. A strike of three Avengers found and attacked *Haguro* that afternoon and scored one near-miss right forward. This was the first dive-bombing attack on a major enemy war vessel in the open sea in the Fleet Air Arm's history. *Haguro* was caught and sunk by five destroyers of the 26th Flotilla in the early hours of the next day.

Ameer, *Emperor*, *Khedive*, *Shah*, *Stalker* and *Empress* took part in various fleet operations along the Burmese coast, over Sumatra, the Nicobars and the Andamans in June and July 1945. On 26 July, when the fleet was carrying out mine-sweeping off Phuket Island, *kamikaze* bombers made two attacks, one in the morning damaging *Ameer* and the cruiser *Sussex*, the second in the evening hitting the mine-sweeper *Vestal* and setting her on fire so that she had to be sunk. This was the last offensive action in the Indian Ocean, and a Hellcat of 896 Squadron which failed to pull out of its dive over the Kra Isthmus was the last aircraft to be lost to enemy action.

By VJ Day, 15 August 1945, the East Indies Fleet had sixteen assault or strike carriers, although some of them arrived too late to see action. Escort carriers covered the unopposed reoccupation of Singapore and the landings at Port Swettenham and Port Dickson in September 1945. In the first seven months of 1945, the fleet's aircraft had accounted for about a third of the Japanese aircraft operating in the Indian Ocean theatre and had destroyed considerable amounts of coastal shipping, shore installations and Japanese transport, besides giving constant air cover to the fleet.

Operation ICEBERG

With one major notch in their gun belts, the British Pacific Fleet arrived in Fremantle on 4 February and in Sydney on the 10th, to a rapturous welcome from the Australians, who had been under 'American occupation' since 1942. After enjoying the fabulous Australian hospitality, the fleet moved up to Manus in the Admiralty Islands, on 7 March 1945 and would soon, it was devoutly hoped, go on to take part in the impending assault on Okinawa (Operation ICEBERG). But at Manus, the fleet waited—and waited. The days passed, and nothing happened. Nobody seemed to want the BPF. They had brought their instruments to the party, 12,000 miles across the world, and nobody was going to ask them to play.

The reason for the delay was the long and complicated political controversy over the fleet's formation and employment. Briefly, the Americans, and particularly Admiral King, the US Naval Chief of Staff, who was no Anglophile, were reluctant to admit the British fleet to the Pacific. It was an American theatre, and King feared that the British would divert supplies and logistical effort away from the US Navy. The BPF was the only flexible reserve force in the Far East and General MacArthur, for his part, wanted it to be used in his south-west Pacific Ocean theatre, against objectives in Borneo and the East Indies. At home the Foreign Office also favoured using the BPF to regain former British possessions in South-East Asia; it was important that such possessions be recovered by British force of arms. Remote islands far off in the Pacific, strategically vital to the Allied cause though they might be, meant nothing to the peoples of Japanese-occupied South-East Asia. In other words, the Japanese must not only be defeated, they must be seen to be defeated. Mr Churchill himself had always favoured the Indian Ocean as the centre of gravity of British operations in the East. He still hankered after an old love of his, often planned for and as often postponed, a combined operation to force the Malacca Strait. However, the Admiralty at home, and Admiral Fraser in the Far East, were both convinced that the main British fleet should be used in the main offensive against Japan, in the Pacific. It was this point of view that Mr Churchill put to President Roosevelt at Quebec in September 1944, offering him the use of a balanced, self-sufficient fleet. The offer was accepted as soon as it was made.

But the decision had not been implemented by the time the BPF reached Manus. So the ships waited, while the temperature rose and spirits dropped. Manus was close to the Equator, and the climate was humid and unhealthy. At midday, eggs could be fried on the flight decks. Ships' companies suffered outbreaks of rashes, prickly heat and tropical skin disorders. Seeadler Harbour was vast and the shore-line almost featureless. There was nothing to do but contemplate suicide. It hardly seemed credible but people actually began to remember the Arctic convoys kindly.

At last, on the morning of 15 March, the glad signal was received ordering the fleet to report to Admiral Nimitz for duty in connection with Operation ICEBERG. There was also a proviso that the fleet was to be at seven days' notice to disengage and be reallocated to the south-west Pacific theatre. But the fleet had eyes only for the main signal and sailed for Ulithi on the 18th, arriving two days later. There, in that giant atoll lagoon, packed with warships and assault shipping, every man in the BPF could at last see for himself what kind of war he was joining.

The fleet sailed for ICEBERG on 22 March, speeded on by a 'good hunting' signal from Admiral Spruance, whose fleet they were to join, being designated as Task Force 57. The BPF's actual strength of two battleships, *King George V* (flag of Admiral Rawlings) and *Howe*, the four carriers *Indomitable* (flag of Admiral Vian), *Illustrious, Victorious* and *Indefatigable*, with 218 aircraft embarked, five cruisers, and eleven destroyers, was actually no stronger than the average American Task Group, but the fleet were given the full title of a Task Force.

For ICEBERG, the fleet had been given the important but secondary task of neutralising, with bombing, fighter sweeps and fighter CAPS overhead, the airfields on the two islands of Miyaka and Ishigaki (names which still haunt the memories of those who served in the BPF),

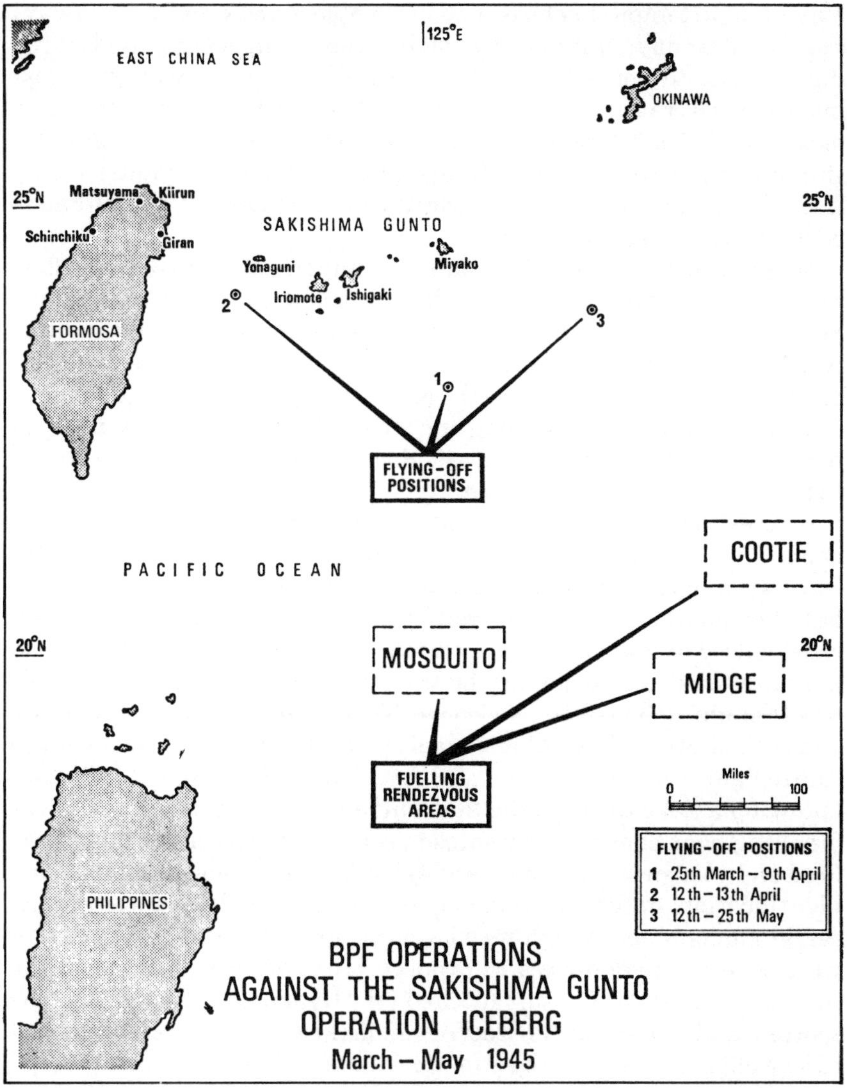

which were part of the Sakishima Gunto, an archipelago stretching like a chain from Formosa north and east up to Okinawa. Task Force 57's duty was to prevent the Japanese staging reinforcement aircraft through the Sakishimas. It was a necessary but thankless task. Off the Sakishima Gunto the fleet stood no chance of glory and every chance of being hammered.

At dawn on 26 March, Task Force 57 was in position some 100 miles south of Miyako, to begin a remarkable operation, unlike anything the

Navy had attempted before. The main objectives were the runways and day after day, in strike after strike, the Avengers trundled off the flight decks to bomb them. By nightfall, the runways were pock-marked with craters and unserviceable. But the Avengers were armed mostly with 1000-lb semi-armour piercing bombs, more suitable for shipping than coral rubble. Overnight the Japanese industriously filled the craters in, so that with the dawning of a new day the job had to be done all over again.

Meanwhile Hellcats and Corsairs accompanied the strikes and flew high CAPs over the targets, Fireflies rocketed coastal shipping, barracks and transport, and the Seafires flew CAPs over the fleet. After two days of strikes, the Task Force retired to a fuelling area to meet a group of tankers and refuel. In the early days there were frequent gear failures and fuelling was a long and frustrating business, taking up to three days, after which, the Task Force returned once again to the flying-off area, for more strikes.

Aircrew losses were steady, rather than spectacular as at Palembang. They amounted to a slow bleeding away of squadron strength. One day an Avenger would succumb to flak over Ishigaki and its crew lost. On another, a Corsair would ditch and though its pilot, an experienced aviator who was Commanding Officer of 1834 Squadron Corsairs, was seen to get out, he was not picked up; later his lifebelt was thought to have been defective. Two Seafires collided in mid-air and both pilots died. Another Corsair took a barrier and went over the side in flames, killing the pilot and three bystanders on deck. A Seafire following a Zeke through the fleet barrage was shot down. A Hellcat breaking back over friendly ships in semi-darkness before dawn was fired on by surprised gunners and the pilot killed. The problem of air identification, differentiating between friendly and Japanese single-seater monoplanes at high speed, was never really resolved and there were several accidents which aroused great bitterness amongst the squadrons. A third Corsair ditched off Miyako and, though he was spotted and reported, he was never picked up. Strenuous air-sea rescue efforts were made by American submarines lying offshore and patrolling 'Dumbo' flying boats, as well as by the fleet's own Walrus, but they were not infallible.

'Bandits' were few but the fighter pilots and the fighter direction officers improved with practice. A flight of Hellcats led by Lt-Cdr Harrington shot down one unwary Oscar by means of a little cunning. Harrington had his Hellcats loaded with a special ammunition 'cocktail', of five ball, one tracer, and one armour-piercing. With this the Hellcats could take any kind of target and the tracer often jolted an enemy into a mistake. 'He was flying out of killing range and I carefully fired a good long blast over his port wing. He very kindly obliged by executing rather a difficult turn to port, which enabled me to close and shoot this unhappy amateur down.'

This 'unhappy amateur' was on his way to attack the fleet in the first of a series of massed suicide attacks in the *kikusui* or 'floating chrysanthemum' campaign with which the Japanese responded to the first landings on Okinawa by the US 10th Army on 1 April. Resistance on shore was slight, the Japanese reserving their resources for later, but at sea, US and British ships were subjected to attacks by more than 300 *kamikaze* aircraft in two days. On the morning of 1 April, one Zeke was engaged over the fleet but evaded pursuit and crashed on *Indefatigable* abreast the island where its 500-lb bomb detonated, denting the flight deck to a depth of three inches and causing extensive damage and casualties in the island superstructure. Lt P.G.H. Roome, RNVR, the Deck Landing Control Officer, described the Zeke's attack in *The Forgotten Fleet*: 'The Jap aircraft came straight for *Indefatigable* from five or six thousand feet and passed her to starboard. It then pulled straight up and looped off the top to come vertically into the ship. I imagine the pilot intended to go down the funnel but missed and hit the flight deck alongside the flight deck sick bay and into one of the barrier stanchions. As it made its first pass ... I decided on discretion and leapt over the port side into the cat-walk and immediately after the bang went back to the flight deck. The remaining bits of aircraft were burning and there was damage to the island sick bay. There was a lot of what I first thought was smoke filling the entrance to the island and Lt-Cdr Pat Chambers, Lieutenant Commander (Flying) emerged bleeding and shocked.' Lt-Cdr Chambers's own recollection in *Task Force 57* is equally vivid: 'I was spun head over heels, over and over, or so it felt, until I got hold of a bit of a jagged cornerpost, by which I heaved myself clear. In the steam and the din there was no sign of my two companions who had been laughing and joking a moment before. There was just a mass of dead and wounded in the area....' The remains of the Japanese aircraft were manhandled over the side. Somebody produced an object for Roome's scrutiny 'which turned out to be a piece of finger. That was all that was found of the pilot.'

The fires were put out, jury barriers rigged, and a Seafire landed on *Indefatigable* only fifty minutes after the attack—an achievement which profoundly impressed the US Navy, in the person of *Indefatigable*'s American liaison officer, who told Captain Graham, 'When a *kamikaze* hits a US carrier it's six months repair at Pearl. In a Limey carrier it's a case of "Sweepers, man your brooms."'

The secret was in the armoured flight decks, but in every other way the British carriers were, as the Americans put it, 'not able to look after themselves', being insufficiently armed with anti-aircraft guns with enough stopping power physically to break up a *kamikaze*'s frame in the air. Several *kamikazes* were hit and flamed during their dives but still completed their attacks.

On the 6th, in another *kikusui*, it was *Illustrious*'s turn. The attacker was only in sight for about ten seconds during which *Illustrious*'s gunners shot away its port wing and part of its tail but the starboard wing tip carried away a radar aerial above the bridge and the aircraft spun into the sea close to on the starboard side abreast the funnel, where its bomb exploded. Parts of the pilot—his skull and both eyeballs—were blown on deck, together with a rubber dinghy—unexpected survival equipment for a suicider. The crew of one pompom were disgusted to find a sliver of burned human flesh, like charred bacon, sticking to their gunsight.

Meanwhile the losses went on. The Japanese gunners, especially on Ishigaki, were alert and uncowed. Two more Avengers were lost. A New Zealand Corsair pilot was shot down making a third and ill-advised strafing run. Another Corsair ditched and the pilot was seen and reported, but not picked up.

Task Force 57 could only listen, with admiration and from a distance, to the reports of the American Task Force 58's dispatch of the giant battleship *Yamato*, making what amounted to a *kamikaze* sortie on a Homeric scale, on 7 April. But on the 12th, in response to an American request, Task Force 57 shifted targets to Formosa where Major Hay was once more in attendance as Air Co-ordinator, in an attack on Matsuyama airfield, in the north of the island. 'From down below I was able to observe the bombs striking the airfield until it was covered with brown smoke and dust. There was no flak before, but the moment the bombers appeared below 3000 feet every gun went into action—they had evidently been waiting for us. On withdrawal one Avenger bombed a factory and after recording the results by camera we strafed what was left. I also caught a passenger train which was skulking in a tunnel but which rather carelessly had left the engine sticking out. On the east coast my Numbers 3 and 4 shot up two junks and we all proceeded to Giran where approximately twelve aircraft of various types were spotted. One twin-engined plane was strafed but did not burn. Considering the weather I think the strike was a fine piece of work. There was no jamming and radio discipline was excellent. No enemy aircraft were seen airborne.'

The same day, two Fireflies of 1770 Squadron, piloted by Lt W. Thomson, RNVR and Sub-Lt W.P. Stott, RNVR, flying CAP for an American Dumbo flying boat, surprised a flight of five Japanese bombers off Formosa and in what Admiral Vian himself called 'a brilliant little encounter' shot down four of them and damaged the fifth.

During a fuelling period on 14 April, *Illustrious* was relieved by *Formidable*. *Illustrious* had difficulty in working up to full speed and still suffered the legacies of her underwater damage in the Mediterranean. She had had a splendid war and it was time for her to go home. *Formidable*'s air group had a speedy initiation to the realities of life off the Sakishima Gunto. On 16 April, Lt-Cdr 'Judy' Garland, Commanding Officer of 1842 Squadron Corsairs, was shot down and killed on his first day over Ishigaki.

On 23 April TF57 arrived at Leyte for a period of rest and recuperation. They had been continuously at sea for thirty-two days, longer than any British fleet since Nelson's day, twenty-six of those days being off an enemy coastline. They had carried out twelve strike days, lost nineteen aircraft to flak and twenty-eight to other causes, had sixteen pilots and thirteen aircrew killed or missing, flown 2444 sorties on strike days, dropped 412 tons of bombs, fired 325 rocket projectiles and had destroyed thirty-three Japanese aircraft in the air and twelve on the ground, not including probables.

Leyte was slightly cooler than Manus but there was no shore leave and the eight days spent there were not a rest cure. 'There was in fact,' wrote Admiral Rawlings in his report, 'little time for rest or relaxation for officers or ratings during this period and after a day or two most of us, I feel sure, wished ourselves back at sea again.'

TF57 sailed from Leyte on 1 May 1945, 'like giants refreshed' as Admiral Rawlings signalled to the fleet train, for another series of strikes. It was the mixture as before, bombing, CAPS and RAMRODS, but there was a new ingredient: a bombardment of Miyako by the heavy guns of the fleet on 4 May. This, as Rawlings and Vian later conceded, was a tactical error, depriving the carriers of anti-aircraft fire. They and their destroyers clustered together, feeling naked and deprived, for the maximum mutual support but they had been sighted by a Japanese shadower and during the day *kamikazes* attacked and hit both *Indomitable* and *Formidable*. After one pass by a *kamikaze*, an officer on *Formidable*'s flight deck, writing in *A Formidable Commission*, saw him come into sight again 'from behind the island, banking hard to close the ship over the starboard quarter. He was still apparently unharmed and now, out astern, the target of fewer guns; for fewer could be brought to bear at that angle, a fact he probably knew. His silhouette changed to a thin line with a bulge in the middle and he seemed to hang in the air as he dived for the ship. I had waited long enough and ran about fifteen yards forward to a hatch, down which I jumped in the company of a rather fat leading seaman. As we hit the deck an immense crash shook the ship. I gave it a second or two to subside, during which the light from above changed to bright orange, and ran up again.

'It was a grim sight. A fire was blazing among wreckage close under the bridge, flames reached up the side of the island and clouds of black smoke billowed far above the ship.'

But within two hours *Formidable* was capable of 24 knots, and four hours later was able to land on her aircraft again. Shortly after the *kamikaze* had crashed, Captain P. Ruck-Keene, in *Formidable*, gripped the arm of his US Liaison Officer, Lt-Cdr 'Ben' Hedges, USNR, and 'in a voice like a fog-horn in the mating season', asked him 'What do you think of our flight decks now?' 'Sir', said Ben, 'they're a honey.'

On 9 May *Formidable* was hit again and *Victorious* who had had a narrow escape, due to Captain Denny's superb ship-handling, on 1 April, was hit twice. Both carriers were able to operate aircraft again, but this long battering was beginning to wear the Task Force down. Experienced pilots were making deck-landing errors which were inconceivable a few weeks before. Lt-Cdr Foster, Commanding Officer of 849 Squadron Avengers, noticed that his crews 'were showing definite signs of strain'. In *Indefatigable*, Captain Graham saw that 'the squadrons first embarked in June and July 1944 were beginning to show signs of strain and tiredness.' There was an epidemic of mumps in the cruiser *Gambia* and of gastro-enteritis in *Indefatigable*. *Formidable* suffered a very serious hangar fire to add to the aircraft losses she had sustained in *kamikaze* attacks. The destroyer *Quilliam*, closing *Indomitable* to act as her 'KK' or counter-*kamikaze* escort, collided with her. The ship's company of the Canadian-manned cruiser *Uganda* were invited by their government to vote on whether they wished to carry on or go home; unsurprisingly, the majority voted to go home.

The last two days of strikes were carried out on 24 and 25 May, when TF57 retired southwards, without giving Ishigaki or Miyako a backward glance. They had been at sea for sixty-two days, broken by eight days at Leyte. They had not sighted an enemy surface vessel at any time, but they had completed another twelve strike days, destroyed another twelve Japanese aircraft, excluding probables, lost another six pilots and two aircrew, lost another seven aircraft to flak and forty-four operationally to ditching, suicide attacks, and deck-landing accidents, flown another 2449 sorties on strike days, dropped another 546 tons of bombs and fired another 632 rocket projectiles. In ICEBERG, TF57 lost twenty-six aircraft to flak and seventy-two operationally but the total loss, from all causes, was 203, or ninety-three per cent of the fleet's aircraft complement of 218 aircraft. These losses, for fifty-seven Japanese aircraft destroyed, suggest a somewhat Pyrrhic victory for the Fleet Air Arm. But aircraft were truly expendable by that stage in the war and TF57 had undoubtedly ground down the enemy's capacity to use the Sakishima airfields and thus contributed to the conquest of Okinawa. Admiral Spruance sent the fleet a gracefully appreciative signal and wrote in his report that the British carrier task force was experienced and competent enough to take its place in the line with the American Fast Carrier Task Force.

Organised Japanese resistance on Okinawa ended on 21 June 1945 and finally ended on 2 July. The Japanese 32nd Army of some 70,000 men was killed or taken prisoner to the last man. The Japanese air force outside Japan had lost nearly 8000 aircraft. The US 10th Army had lost 7200 dead or missing, with another 31,000 wounded. The US Navy had lost 4900 killed or missing, and 4800 wounded, 540 aircraft, thirty-six ships sunk and 368 damaged (the great majority to *kamikaze* attack). The Royal Navy had eighty-five killed or missing, and eighty-three wounded. All five carriers had been hit by *kamikazes*, and the destroyer *Ulster* so badly damaged that she had to be withdrawn.

The BPF had recently been reinforced by the arrival of the new carrier *Implacable* which had, certainly for the BPF at least, the impressive complement of seventy-eight aircraft—forty-eight Seafires of 801 and 880 Squadrons, twelve Fireflies of 1771, and twenty-one Avengers of 828. Largely through the ingenuity and some bribery of her Commander (Air) C.L.G. Evans (the same Evans whose squadron had downed the first German aeroplane of the war in 1939), *Implacable* had solved the short-range problem of the Seafires by obtaining and fitting American ex-Kittyhawk drop-tanks obtained by subterfuge from a dump in New Guinea. With these, as Evans said, 'We flew Seafires on every strike and they made superb fighter-bombers'.

Their first baptism of fire was on Truk, the great Japanese stronghold in the Carolines. *Implacable* and *Ruler* (twenty-four Hellcats of 885), escorted by three cruisers and five destroyers sailed from Manus on 12 June, their main object being to give newly arrived aircrew some action experience although the training aspect was not stressed. As Evans said, 'I never told my pilots it was a training expedition; if you say that, you start losing a few people.'

The force arrived off Truk at dawn on the 14th. Avengers, Seafires and Fireflies bombed and strafed targets ashore. Avengers bombed by the light of flares that night and in the morning the cruisers carried out bombardments. Two aircraft were lost but the harm done to Truk was not great. The fact was that Truk had already been pounded so hard and often by the Fast Carrier Task Force that it was no longer a major objective.

Operations off Japan

For all their exploits against *Bismarck* and *Tirpitz*, for all the campaigns they fought in, from Norway to North Africa, for all their successes from Salerno to Sumatra, the brief operations (only eight strike days) of Task Force 37 off the coast of Japan in July and August 1945 remain the Fleet Air Arm's most polished professional performance of the war. Though always short of tankers, and with replenishment a constant worry, the British Pacific Fleet operated in close company with the US Third Fleet and struck blow for blow, ship for ship, almost until the final surrender of Japan. Only shortage of fuel robbed some of the ships of their proper place at the end.

The fleet sailed from Sydney on 28 June 1945, very much refreshed after three weeks in harbour. There were replacement aircraft on board, and fresh faces in the ready rooms. The battle-fatigued aircrews of ICEBERG had been relieved—and not before time, as they themselves had complained. *Indomitable* was being refitted, so Vian shifted his flag to *Formidable* which had thirty-six Corsairs of 1841 and 1842, twelve Avengers of 848, and six Hellcats from *Indomitable*'s 1844 Squadron for photo-reconnaissance and night-fighting. *Victorious* had thirty-seven Corsairs of her old friends 1834 and 1836, with eighteen Avengers of 849, and *Indefatigable* (who was delayed by machinery defects and sailed later) had forty-eight Seafires of 887 and 894, twelve Fireflies of 1772, and twenty-one Avengers of 820. With *Implacable*'s seventy-eight aircraft (strength as in the Truk strike) the fleet (now redesignated Task Force 37, Halsey having relieved Spruance and the Fifth Fleet having become the Third) had a total aircraft complement of 255—the most any Fleet Air Arm force had ever put in the field at one time.

The American carriers of Task Force 38 had already begun operations on 10 July with preliminary strikes against airfields on the Tokyo plain and had then moved north to attack northern Honshu and Hokkaido on 13, 14 and 15 July. Some hundreds of Japanese aircraft were destroyed on the ground, but air opposition was almost non-existent. The Japanese were husbanding their air strength against the time which they knew would come, when the invaders attempted to set foot on sacred Japanese soil.

The US Third Fleet had nine fleet carriers, six light fleet carriers, seven battleships, fifteen cruisers and sixty destroyers, deployed in three Task Groups. This vast fleet was refuelling, forming in Admiral Rawlings's words 'a striking and unforgettable picture' when the British Pacific Fleet, with three carriers, one battleship *King George V* (flag of Admiral Rawlings), six cruisers and fifteen destroyers joined

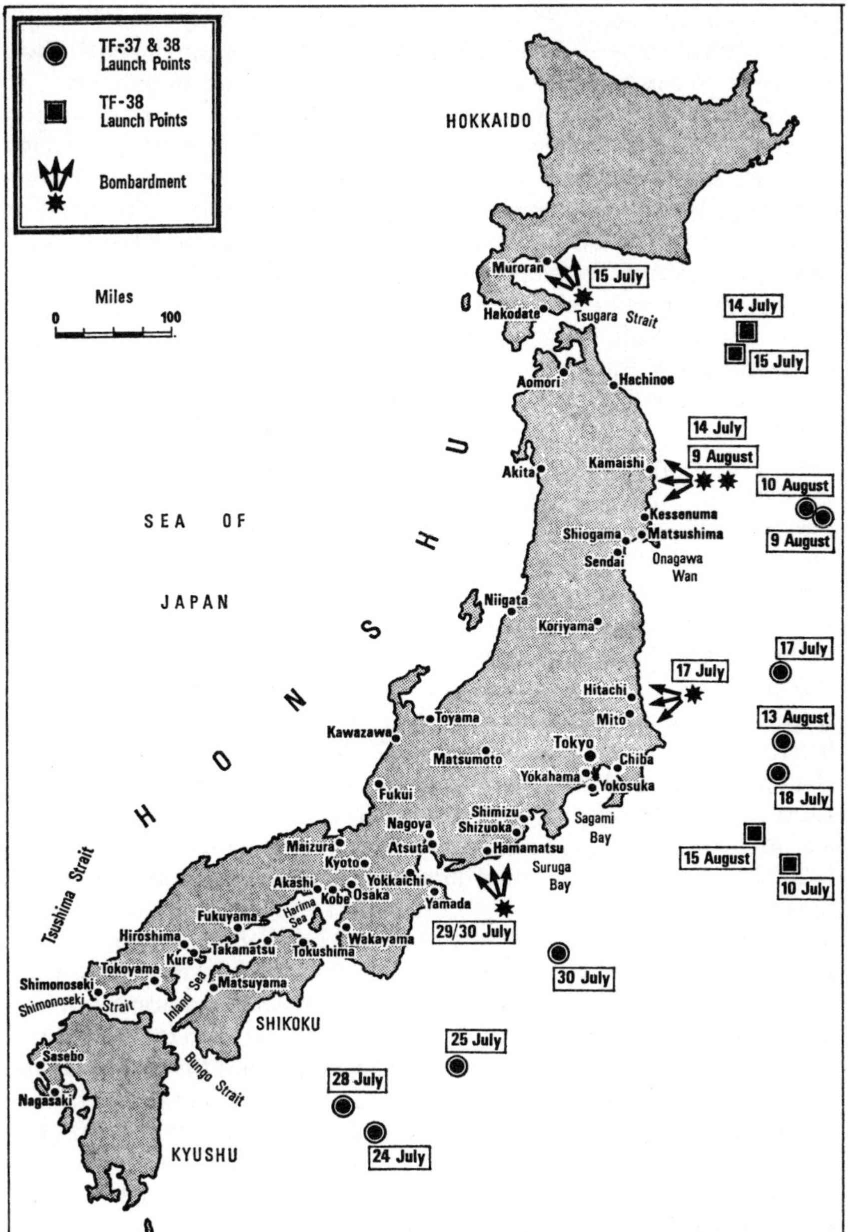

OPERATIONS AGAINST THE JAPANESE HOME ISLANDS
July and August 1945

early on 16 July, some 300 miles east of Japan. Rawlings and Vian and their staffs at once repaired on board Halsey's flagship, the battleship *Missouri*, to pay their respects and meet their fleet commander.

As always with the BPF, tactics had to be disentangled from politics. Nimitz, possibly under political pressure, had countermanded Halsey's expressed wish to use the British force as part of his own (although Fraser had no objection at all to his ships being used as such). Halsey offered Rawlings the alternatives of operating in all respects except name as part of the Third Fleet, or operating semi-independently some sixty miles away, or operating entirely separately against 'soft' targets which Halsey's staff would recommend. Unhesitatingly Rawlings chose the first option and as Halsey said, 'my admiration for him began at that moment'. Officially it was intimated that the British fleet would operate sixty miles apart but in fact the Task Groups were fifteen miles across. Sixty miles was therefore the distance across the van of the fleet and placed the British Task Force neatly, and proudly, in the honoured station on the right of the line. Admiral Fraser, though quite content with the outcome, later commented of this anomaly, 'Provided he obeys the letter of the law, even if he completely disregards its spirit, every American is quite happy that the right and sensible action has been taken.'

And so, at 4 p.m. on 16 July, the combined fleets moved off towards the first flying-off position in the order from north to south, Task Force 37 (Rawlings), Task Group 38–1 (Rear Admiral T.L. Sprague), Task Group 38–4 (Rear Admiral A.W. Radford), and Task Group 38–3 (Rear Admiral G.F. Bogan) with the Task Force Commander Vice Admiral J.S. McCain, flying his flag in the fleet carrier *Shangri La*.

TF37's targets for the first day, 17 July, were airfields at Sendai, Masuda and Matsushima. The first strikes on Japanese soil by British aircraft, a RAMROD of sixteen Corsairs from *Formidable* and seven Fireflies from *Implacable*, were flown off from a position 250 miles east-north-east of Tokyo. Later in the day, another RAMROD of ten Corsairs from *Formidable* and twelve from *Victorious*, led by Lt-Cdr J.G. Baldwin, Commanding Officer of 1834 Squadron, flew right across Japan to the west coast and attacked targets at Niigata, and strafed shipping in the Sea of Japan—again the first appearance of British aircraft there. It was a day of 'firsts' because, later that night, TF 37 flew a night CAP (for the first time) over a bombardment force which included *King George V*, making the first bombardment of Japan by a British capital ship.

On the days of 18, 24, 25 and 28 July, Task Force 38 completed the demolition of the remnants of the Japanese Navy. The surviving major Japanese warships were all permanently moored alongside, for lack of fuel, heavily camouflaged and acting as no more than floating anti-aircraft batteries. American aircraft sought them all out, with bombs and torpedoes, and sank or heavily damaged the battleship *Nagato* at Yokosuka, the battleships *Hyuga*, *Isa* and *Haruna*, the cruisers *Tone* and *Aoba*, the aircraft carrier *Amagi* and several cruisers and destroyers at Kure, on the Inland Sea.

The British carriers were excluded from these strikes, being given secondary targets at airfields and towns nearby. The reason was ostensibly the short range of the Seafire, which even the Americans admitted was a specious pretext. The real reason was that the US Navy wanted full and final revenge for Pearl Harbour and did not wish to share the execution with outsiders. As Admiral Halsey admitted after the war, he realised that his action was ungenerous and unworthy, but he had yielded to the pressure of his staff, who insisted that he should forestall any possible post-war claim by the British that they had delivered even a part of the final blow which demolished the Japanese Navy. It was a churlish way to treat an Ally who had declared war on Japan only a few hours after the attack on Pearl Harbour.

However, on 24 July, British aircraft did come across a Japanese carrier in Shido Wan, a bay on the east coast of Shikoku, the *Kobe* Class light escort carrier *Kaiyo*. In several strikes, Avengers, Fireflies, Corsairs and Seafires, including aircraft from all four carriers, attacked *Kaiyo* and left her badly damaged, with a broken back. This, surprisingly, was the only attack by Fleet Air Arm aircraft on an enemy carrier during the whole war (although *Kaiyo* was also attacked the same day, and claimed, by American aircraft).

In general TF37 conformed to American routine and practice. Vian took over tactical control of the fleet from Rawlings whenever flying was in progress. TF37 observed 'point option'—an ingenious scheme for keeping a large fleet together. An imaginary point was presumed to move across the chart at a known course and speed; ships and task groups could manœuvre individually at will, provided in general they kept station on point option's course and speed. Occasionally, however, TF37 differed, as on 18 July when the targets had been airfields and installations in the Tokyo area. *Implacable* frequently had more aircraft in the air than other carriers and when the fleet came to retire, had still not landed on the last dozen of her Seafires. Mist was closing in. Vian was anxious to get away. Grimly ignoring ever more insistent signals from his force commander, Captain Hughes Hallett kept *Implacable* standing on towards the Japanese coastline, urged by his Commander (Air). Eventually the twelve Seafires were landed on, when *Implacable* was less than sixty miles offshore, almost in the opening of Tokyo Bay. As Evans said, 'Vian never forgave me for that.'

But Vian himself smarted under the criticism that TF37 had become separated by 100 miles from the American ships that day 'owing to the inexperience of our Carrier Admiral. This hurt.'

The Seafires, now used as ground-strafing aircraft so frequently that even Vian began to soften towards them, had developed their own tactics for airfield attacks, as Lt-Cdr M. Crosley, Commanding Officer of 880 Squadron Seafires explained in *The Forgotten Fleet*: 'As we were mostly over the sea, it was the inlets and the green terracing on the hillsides which we noticed first, and the utter peacefulness of it all, except when we went near an airfield, when the scene changed somewhat—I have never seen quite so much 20-mm and 40-mm flak! The general sensation of being over Japan was one of foreboding, deep fear. We had heard tales of what the locals did to airmen who got hacked down. We got "in and out" as quickly as we could. We used the "low-pull up-low" pattern which is still done today, approaching the target area at zero feet all the way in.

'We kept below the radar echo height as long as possible so that we did not give warning of our approach. Near the target we split up and climbed to about 8000 feet and approached in a sort of "scissors", with aircraft attacking from different angles. Thus we gave the gunners no time at all to get their fingers out before we were gone—not to return. Occasionally, when we had twelve or even sixteen aircraft on a raid (all Seafires on air-to-ground strafing) we would all cross over the middle of the enemy airfield. The main hazard was mid-air collision or ricochets from our own bullets (2 × 20mm, plus 4 × .303s). The angle of dive was as steep (45–60°) as possible (out of the sun if any) from 8000 to 6000 feet. We then made our exit at zero feet.'

Air opposition was so slight that there were only two interception incidents during TF37's period off Japan. But one of them, on the night of 25 July, showed once again that the price of safety in war was eternal vigilance. The fleets were withdrawing to the fuelling area when a shadower located them and the enemy made a serious attempt to attack. The fleet fighter direction officer, Cdr E.D.G. Lewin (the same man who had flown *Ajax*'s Seafox at the River Plate nearly six years earlier) noticed a small group of 'bogeys' on the radar screens, bearing north-east distance ninety-four miles. TF37 had only four dusk patrol Hellcats from *Formidable* airborne at the time and, despite the scepticism of the direction officers in the nearest American task group, Lewin directed the Hellcats on to the intruders and intercepted them at 20,000 feet, thirty miles from TF37. They proved to be four Japanese 'Grace' Aichi torpedo-bombers, heading for an attack on the fleet. The Hellcats shot down three and damaged the fourth. Now thoroughly convinced, the American direction officers guided fighters from the night carrier *Bon Homme Richard*, which with gunfire from picket destroyers, broke up another larger group of intruders and shot down one of them.

The weather proved a more awkward opponent than the Japanese. It was the typhoon season and several days' strikes were cancelled because visibility deteriorated. Maddeningly, it was either often clear inland, when the ships were hidden in thick fog, or strikes were flown off in the clear, only to find targets inland shrouded in mists. Skirting round the edges of typhoons, the fleets still encountered gigantic swells which made fuelling and transferring stores particularly difficult.

Task Force 112, the Fleet Train, was one of the most extraordinarily motley collection of ships ever assembled in the Navy's history. In contrast to the ships of the US Navy's Fleet Train, which were all commissioned as warships under naval discipline, the British fleet train had Norwegian masters, Chinese deck-hands, Dutch mates, Lascar firemen and Papuan winchmen. There were officers and men from all over the Commonwealth, Australia, New Zealand, South Africa, Canada, India and the West Indies. There were warships flying the White Ensign, Royal Fleet Auxiliaries flying the Blue, and chartered merchantmen flying the Red. There was a Panamanian collier, and a Dutch hospital ship, and a Danish tanker, and a Belgian ammunition ship. There were net layers and accommodation ships and naval stores' ships and radar/radio maintenance ships.

Replacement aircraft for the fleet were carried by the 30th Aircraft Carrier Squadron (Cdre R.P. Carne, flying his broad pennant in *Striker*). *Striker*, *Slinger*, *Speaker*, *Chaser* and *Arbiter* were all engaged in ferrying duties, taking fresh aircraft up to the striking areas and bringing back flyable 'duds'. Fresh aircrew were also taken up to the operating zone, keeping in flying practice on the way. *Ruler*'s Hellcats frequently provided CAPS over the fuelling areas, to give the fleet carriers' air groups some rest. Ashore in Australia, Mobile Naval Air Bases (MONABS) were set up at Nowra, Bankstown, Schofields, Jervis Bay and Maryborough. The Americans also turned over to the Fleet Train a complete air strip at Ponam Island, Manus.

The fleet's air groups were generally in much better physical and mental health off Japan than they had been in the latter days of ICEBERG. There was a much greater choice of targets and missions than on what one Captain called 'the long, drawn-out, stale and arid target features of the Sakishima Gunto'. All the same, there was still a steady drain on squadron strength and in fact, twenty-six pilots, four observers and two TAGS were killed or reported missing during July and August. 'One saw an enormous number of one's friends killed, or go missing, or just not come back', said one pilot in *Task Force 57*. 'One literally had a carrier full of replacement crews following the fleet around and every time the ship replenished this carrier would ferry fresh crews aboard and take off the wounded. One got the feeling that life was a very short term proposition.'

On 9 August, *Formidable* lost one of the most notable personalities in her air group, Lt Robert Hampton Gray, RCNVR, Senior Pilot of 1841 Squadron Corsairs. Gray won a DSC for his services in ICEBERG and was one of the most aggressive and skilful pilots in the fleet. On the morning of the 9th he led the second of the day's RAMRODS, of eight Corsairs, on an anti-shipping sweep along the east coast of Japan. As his RAMROD was crossing the bay of Onagawa Wan, they were fired on by an old Japanese destroyer lying close inshore and camouflaged with anti-aircraft batteries around it. Gray's Corsair was carrying a 1000-lb bomb, which most pilots naturally wished to get rid of as quickly as possible to leave their aircraft more manœuvrable. Gray dived to attack, flying low to ensure success and though his bomb hit and sank the destroyer, Gray himself was shot down and killed. On 12 November 1945, Gray was awarded a posthumous Victoria Cross, his citation mentioning his 'great valour, brilliant fighting spirit and most inspiring leadership'.

But at the time, his squadron Commanding Officer, Lt-Cdr R.L. Bigg-Wither, RNVR, had mixed feelings. 'It has always seemed such a terrible shame to me that this quite unexpected chance should have occurred when the war was virtually over,' he said in *The Forgotten Fleet*. For the war was almost over. Gray's VC was in a sense one of the saddest ever awarded. The cause for which he gave his life was already won. On 6 August Halsey was ordered to withdraw his ships from the operational area for a 'special operation'. Next day, the fleet heard the news of the atomic bomb dropped on Hiroshima. For the first time, the men of the fleet lifted their heads from the business of war and sniffed the first hint of peace. The second atomic bomb on Nagasaki seemed to clinch it and, although the fleet went on striking on 9 and 10 August, the end could not now be far off.

Unhappily, most of TF37 were not there to see it. Shortage of fuel forced the fleet to retire, leaving behind a token force of *King George V*, *Indefatigable*, two cruisers and ten destroyers. The retiring carriers steamed south, knowing they had done well. They had flown 2615 sorties on strike days, destroyed 347 enemy aircraft, and 356,000 tons of shipping. Ship for ship they had kept up the same rate of striking as the Americans. They themselves had lost 101 aircraft, thirty-nine of them to flak.

Indefatigable's air group carried out two more strike days, on 13 and

15 August, after the rest had gone. The last Fleet Air Arm pilot to be shot down in the war was Sub-Lt F. Hockley, RNVR, of 894 Squadron, escorting a strike on the morning of 15 August. His Seafire was hit but he baled out and landed safely on the Chiba peninsula, east of Tokyo. He was captured and taken to the nearest Japanese army headquarters. Later that same day, Hockley was led out and shot. Three Japanese officers were convicted of his murder—two were executed and the third sentenced to fifteen years' imprisonment.

In the days leading up to the Japanese surrender, it was appropriate that the carriers held the ring, as they had so often done, staying at sea to guard against possible last-minute treachery. When, on 2 September, the documents of the formal Japanese surrender were signed on board the US battleship *Missouri*, a thousand Allied aircraft flew over Tokyo Bay.

They made a fine and a brave sight, but the clear air beyond their ranks was full of ghosts, of Esmonde and Gray and the men who flew with them, of Bill Lucy and the resourceful Mid Gallagher and all the crews lost in Norway and the Arctic, of Bayly at Taranto, and Dalyell-Stead off Matapan, and Maund (missing on a flight from Malta in January 1943), of Dickie Cork, of the two gallant jockeys Furlong and Everett, of Sub-Lt Hockley and the nine men murdered after Palembang, of all the hundreds of young men, the pilots, observers and air-gunners, who flew out over the sea and never came back. At times aircrew losses were so high that fellow ship's officers were afraid to befriend them. 'When you went missing that night,' a friend in *Victorious* told John Hoare after his *Bismarck* episode, 'I swore I'd never again make a friend of a flying type. You could go on for ever worrying each time they took off and breaking your bloody heart when they didn't come back.'

But they are all still remembered, in the best of ways and by the best of men, by those who flew alongside them. 'At the Memorial Service at Lee-on-Solent,' wrote Nat Gold, 'I try not only to remember the dead, but the living, the pilots who brought me safely back, the wonderful comradeship—and appreciating how sweet life *really* is before going into action.

'Let us pray it never happens again.'

Index